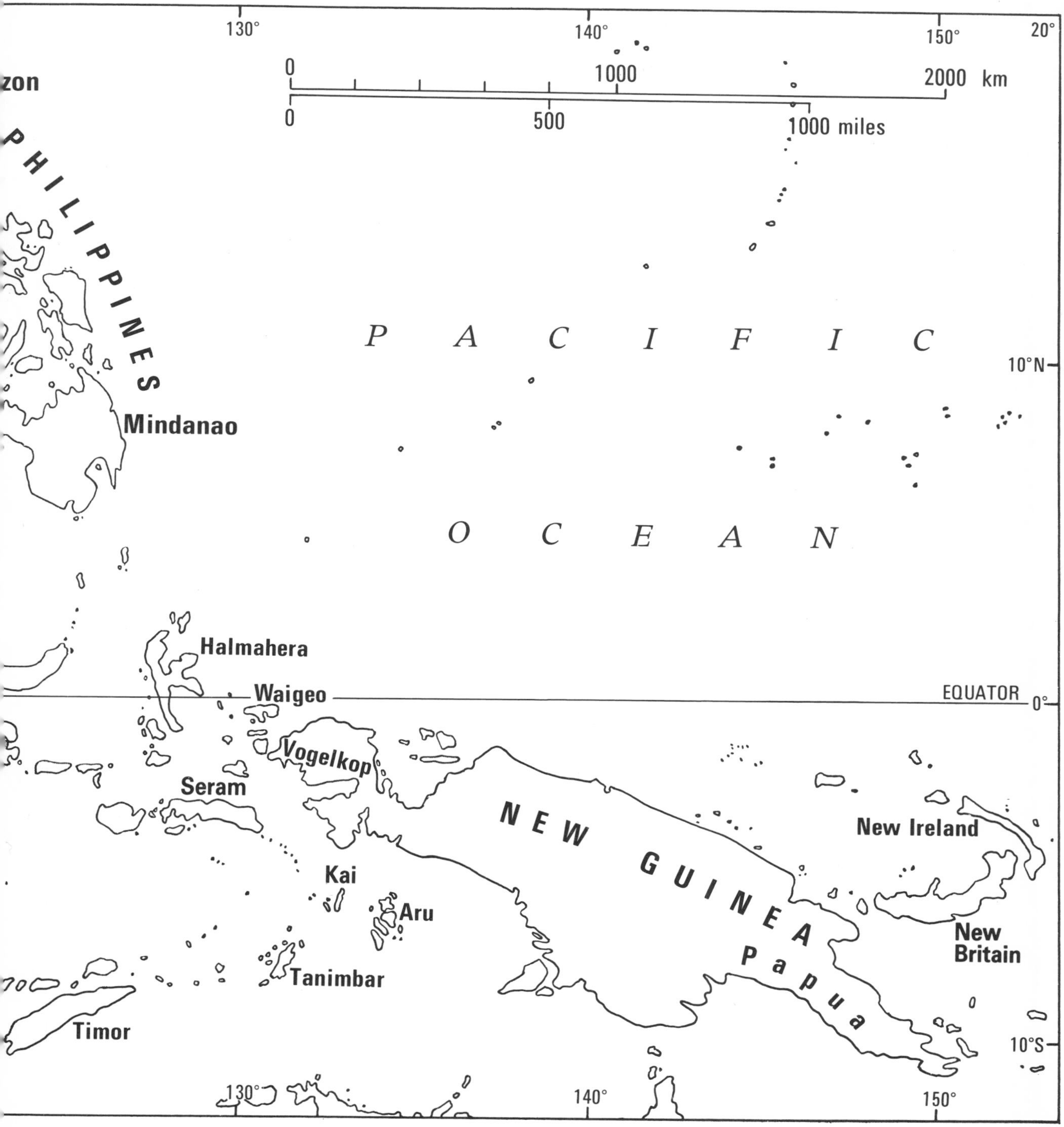

130° 140° 150° 20°

0 1000 2000 km

0 500 1000 miles

zon

PHILIPPINES

Mindanao

P A C I F I C

O C E A N

10°N

Halmahera

Waigeo

EQUATOR 0°

Vogelkop

Seram

NEW GUINEA

Papua

New Ireland

New Britain

Kai

Aru

Tanimbar

Timor

10°S

130° 140° 150°

OXFORD MONOGRAPHS ON BIOGEOGRAPHY

Editors: W. GEORGE, A. HALLAM, AND T. C. WHITMORE

OXFORD MONOGRAPHS ON BIOGEOGRAPHY

Editors

Wilma George, Department of Zoology, University of Oxford.

A. Hallam, Department of Geological Sciences, University of Birmingham.

T. C. Whitmore, Oxford Forestry Institute, University of Oxford.

In an area of rapid change, this series of Oxford monographs reflects the impact on biogeographical studies of advanced techniques of data analysis. The subject is being revolutionized by radioisotope dating and pollen analysis, plate tectonics and population models, biochemical genetics and fossil ecology, cladistics and karyology, and spatial classification analyses. For both specialist and non-specialist, the Oxford Monographs on Biogeography will provide dynamic syntheses of the new developments.

1. T. C. Whitmore (ed.): *Wallace's line and plate tectonics*
2. Christopher J. Humphries and Lynne R. Parenti: *Cladistic biogeography*
3. T. C. Whitmore and G. T. Prance (ed.): *Biogeography and Quaternary history in tropical America*
4. T. C. Whitmore (ed.): *Biogeographical evolution of the Malay Archipelago*

Biogeographical Evolution of the Malay Archipelago

Edited by

T. C. WHITMORE

University Senior Research Officer,
Oxford Forestry Institute,
Department of Plant Sciences,
Oxford

CLARENDON PRESS · OXFORD
1987

Randall Library UNC-W

Oxford University Press, Walton Street, Oxford OX2 6DP

Oxford New York Toronto
Delhi Bombay Calcutta Madras Karachi
Petaling Jaya Singapore Hong Kong Tokyo
Nairobi Dar es Salaam Cape Town
Melbourne Auckland

and associated companies in
Beirut Berlin Ibadan Nicosia

Oxford is a trade mark of Oxford University Press

Published in the United States
by Oxford University Press, New York

© *The contributors listed on p. ix, 1987*

All rights reserved. No part of this publication may be reproduced, stored in a retrieval system, or transmitted, in any form or by any means, electronic, mechanical, photocopying, recording, or otherwise, without the prior permission of Oxford University Press

British Library Cataloguing in Publication Data

Biogeographical evolution of the Malay
Archipelago.—(Oxford monographs on
biogeography; no. 4)
1. Biogeography—Malay Archipelago
I. Whitmore, T.C.
574.9598 QH186

ISBN 0–19–854185–6

Library of Congress Cataloging in Publication Data

Biogeographical evolution of the Malay Archipelago.
(Oxford monographs on biogeography; no. 4)
Based on papers presented at a symposium at the Third
International Congress of Systematic and Evolutionary
Biology held at Brighton, UK, in July 1985.
Bibliography: p.
Includes Index.
1. Paleobotany—Cenozoic—Congresses. 2. Paleontology—
Cenozoic—Congresses. 3. Paleobotany—Malay Archipelago—
Congresses. 4. Paleontology—Malay Archipelago—
Congresses. 5. Paleogeography—Malay Archipelago—
Congresses. I. Whitmore, T. C. (Timothy Charles)
II. International Congress of Systematic and Evolutionary
Biology (3rd: 1985: Brighton, East Sussex) III. Series.
QE926.B56 1987 560'.1'7809598 86—28481

ISBN 0–19–854185–6

Set by Wyvern Typesetting Limited, Bristol

Printed by St Edmundsbury Press,
Bury St Edmunds, Suffolk

QE
926
.B56
1987

PREFACE

Biogeography has been revolutionized by the discoveries of plate tectonics. In 1981 the first in this series of monographs described the origin of the Malay Archipelago by the collision of converging outliers of Laurasia and Gondwanaland in the Tertiary. The major zoogeographical boundary known as Wallace's Line was seen to be a consequence of historical geography. A few biogeographic distributions were difficult to explain on this model and appeared to demand earlier interchange between Laurasia and Gondwanaland.

It was not long before further geological research led to the discovery of Gondwanic terranes embedded in present-day south and east Asia, which broke off the Australian part of Gondwanaland and drifted north earlier than the Tertiary. These important geological discoveries and new evidence on past climates (which has recently become available for the Tertiary) were presented at a Symposium held at the Third International Congress of Systematic and Evolutionary Biology, held at Brighton in July 1985. The revised historical background was then used to re-explore aspects of the biogeography of the archipelago and to consider also the early history of flowering plants in Australia, about which there have been considerable recent advances.

This volume of essays represents the state of knowledge on Malesian biogeography in 1985, much advanced on knowledge only four years earlier. The picture is clearer than that painted in 1981, but new data have thrown up new problems and these are pinpointed to serve as foci for future attention.

We express our thanks to the organizers of ICSEB III, and through them especially to the Royal Society and British Council for financial support. Secretarial assistance with arranging the Symposium and then with typing, and retyping the draft chapters is very gratefully acknowledged to Morley Mansfield, Marion Bakker, and Caroline Budden.

Oxford
May 1986

T. C. W.

CONTENTS

CONTRIBUTORS

M. G. Audley-Charles, Department of Geological Sciences, University College London, Gower Street, London, UK.

M. M. J. van Balgooy, Rijksherbarium, Schelpenkade 6, 2300 RA Leiden, The Netherlands.

J. Dransfield, Herbarium, Royal Botanic Gardens, Kew, Richmond, Surrey, UK.

J. Flenley, Geography Department, University of Hull, Kingston upon Hull, UK.

W. George, Zoology Department, University of Oxford, South Parks Road, Oxford, UK.

J. D. Holloway, Commonwealth Institute of Entomology, 56 Queen's Gate, London, UK.

A. P. Kershaw, Department of Geography, Monash University, Clayton, Victoria 3168, Australia.

R. Morley, Robertson Research International Ltd, Pen y Coed, Llandudno, Gwynned, UK.

G. G. Musser, American Museum of Natural History, Central Park West at 79th Street, New York 10024, USA.

I. R. Sluiter, Paleoservices Ltd, Paramount Industrial Estate, Sandown Road, Watford, UK.

A. Takhtajan, Komarov Botanical Institute, Leningrad,USSR.

E. M. Truswell, Bureau of Mineral Resources, Canberra 2601, Australia.

T. C. Whitmore, Oxford Forestry Institute, Department of Plant Sciences, University of Oxford, South Parks Road, Oxford, UK.

NOTE ON PLACE NAMES AND UNITS

In general we follow *The Times Atlas of the World, Comprehensive Edition* (5th edition, revised 1977). Place names mentioned in the text are shown on the end-paper map and on the map of central Malesia, Fig. 1.1. All geological strata referred in the text are shown and dated on Table 1.1.

Banda Arc. The geological term for the double arc of islands from Flores east through Alor and Wetar and north to Banda (Inner Banda Arc) and from Raijua through Timor and the Tanimbar islands then north through the Kai islands and west to Seram and Buru (Outer Banda Arc) (see Fig. 1.1).
Cambodia. Kampuchea.
Greater Sunda Islands. Borneo, Java, and Sumatra.
Indochina. Present-day Kampuchea, Laos, and Vietnam.
Inner Banda Arc. See Banda Arc.
Irian Jaya. The western half of New Guinea. Part of Indonesia.
Lesser Sunda Islands (Nusa Tenggara). The geographical term for the islands east of Java from Bali and Lombok eastwards to Damar and Babar (see Fig. 1.1).
Ma. The megayear, 1 000 000 years.
Malaysia. The political state comprising Peninsular Malaysia together with Sabah and Sarawak in northern Borneo.
Maluku. Moluccas (q.v.).
Malesia. The biogeographical province stretching from Sumatra and the Malay peninsula south of the Kangar–Pattani line (Whitmore 1975) to the Bismarck archipelago.
Moluccas. The geographical term for the islands which occupy the region between Sulawesi (plus the Talaud and Sula islands, Buton and the Tukangbesi islands), the Lesser Sunda Islands (q.v.), Aru and New Guinea (plus Misool and Waigeo). The biggest Moluccan islands are Halmahera, Seram, Buru, and Tanimbar (see Fig. 1.1).
Nusa Tenggara. Lesser Sunda Islands (q.v.).
Pangaea. A single supercontinent which existed from the Permian until it began to break up in the early Mesozoic.
Outer Banda Arc. See Banda Arc.
Papuasia. New Guinea, the Bismarck archipelago, and the Solomon archipelago.
Sulawesi. Celebes.
Sundaic. Pertaining to Sundaland.
Sundaland. The lands of the Sunda continental shelf, west of Wallace's line.
Sunda shelf. See Sundaland.
Vanuatu. New Hebrides.
West Gondwanaland. That part of Gondwanaland which today forms Africa and South America.

1 INTRODUCTION

T. C. Whitmore

The first volume in this series of monographs, *Wallace's line and plate tectonics*, was published in 1981. In two essays the Malay archipelago (Malesia: see end-paper map and Fig. 1.1) was shown to have been created by collision during the Tertiary, from 15 Ma (mid-Miocene) to 3 Ma ago, at about the position of Sulawesi, of outliers of Laurasia and Gondwanaland, as these two supercontinents converged. Much of the Quaternary experienced cooler, drier, more seasonal climates than today. Other essays in the volume showed that if allowance is made for the influence of past climates on present-day ranges, then the biogeography of vertebrate animals, palms, and certain other plants can be seen to reflect the geological history. Wallace's line, one of the world's sharpest zoogeographical boundaries, was seen to be a consequence of historical geography (embracing plate tectonics and climatic fluctuations). A few distributions remained which are difficult to explain on this model. Most puzzling perhaps are Fagaceae (the beeches and oaks), which are a northern hemisphere family of flowering plants, reaching south-eastwards across the Equator into New Guinea, but with a southern temperate genus, *Nothofagus*, the southern beeches, which extend northwards into the tropics, also in New Guinea. Some

Fig. 1.1. The central part of the Malay archipelago to show the Lesser Sunda Islands (Nusa Tenggara), Moluccas (Maluku) and the Banda Arcs.

pairs of flowering plant families have a northern and southern hemisphere member, extending into the tropics and overlapping in the Malay archipelago. Magnoliaceae (northern) and Winteraceae (southern) are a good example.

The 1981 analysis, explaining present-day biogeography and Wallace's line in terms of a mid-Miocene collision between Laurasia and Gondwanaland, ended by discussion of the remaining enigma these puzzling plant families represent. A hint existed that a shard of Gondwanaland had split off Australia and drifted north, perhaps in late Jurassic or early Cretaceous time. More detail was awaited, as were more data on Tertiary climates and on the evolution and migration of plants and animals, especially additional fossil evidence. Now, only five years later, we have new geological discoveries which might help to explain the enigma.

Mesozoic plate tectonic history now shows (Table 1.1, and Audley-Charles, Chapter 2) southeast Laurasia to contain, in Burma, Thailand, Sumatra, and Malaya several embedded terranes which are continental fragments rifted from Australia–New Guinea, i.e. the north margin of east Gondwanaland, perhaps as late as Jurassic Oxfordian time (160 Ma). These provided stepping-stones of dry land during the Tertiary before the previously described mid-Miocene collision. No longer do biological reconstructions need to accommodate a vast empty stretch of ocean between Asia and Australasia until 15 Ma.

In the present book, several important implications of the new palaeogeographical model are explored. The place or places of origin and diversification of the flowering plants still remains mysterious and controversial. Despite many discoveries of both macrofossils and fossil pollen the palaeontological record is still inadequate, and we must still argue from phylogenetic geography. In Chapter 3 Takhtajan reviews the evidence. The greatest concentration of archaic angiosperms is in east Gondwanaland, pointing to their cradle lying 'somewhere between Assam and Fiji'. He believes this must have lain on fragments, rifted off Gondwanaland; a mountainous archipelago would provide excellent conditions for the survival of macromutant 'hopeful monsters' and for subsequent adaptive radiation. The fossil record of angiosperms in Australia (Truswell, Kershaw and Sluiter, Chapter 4) does not support Takhtajan; the first angiosperms appear there 10 Ma more recently than elsewhere in the world, and unless the fossil record is incomplete this argues against origin and diversification of the group on an archipelago off the northern coast. There is however evidence of floral exchange in both directions between Australia and regions to the north in the late Cretaceous and early Tertiary, before the mid-Miocene collision, though no evidence in the fossil record of a massive influx into Australia after the collision.

Present-day ranges are partially dependent on past climates, for example the disjunct distributions of plants of the highest mountain peaks and others of seasonal (monsoon) climates. Morley and Flenley (Chapter 5) review evidence for the interrelated changes in sea level, in degree of seasonality and precipitation, and in variations of temperature. The mechanisms believed to drive these changes are described. For the first time the fluctuations long known to have occurred during the Quaternary are shown to have occurred also in the late Tertiary. New evidence shows that the Malay archipelago has had extensive areas of seasonal forests (just as has perhumid Africa and America (Whitmore and Prance 1987)). The evidence includes the first record of *Pinus* for Malaya, probably growing as in Thailand in woodlands over grass.

New taxonomic analysis of the palms (Dransfield, Chapter 6) now points away from the previous belief of their origin in south-west Gondwanaland, to an earlier origin in Pangaea followed by diversification of northern Laurasian and southern Gondwanic stocks. Malesian palms come from both stocks. The ranges of some within the archipelago still remain difficult to understand, but most can be well explained in terms of the new palaeogeographical model.

Table 1.1

Stratigraphical ages: major geological events and evidence of the early angiosperms (after Harland *et al.* 1982)

Million years	Era	Period	Epoch	Stage	Events
2.0	CENOZOIC	Q	Pleistocene*		
5.1		TERTIARY (Neogene)	Pliocene		
8.2			Miocene	Messinian	(7)
24.6		(Paleogene)	Oligocene		
38			Eocene		(6)
54.9			Paleocene		
65	MESOZOIC	CRETACEOUS	Upper	Senonian: Maastrichtian	
73				Campanian	
83				Santonian	
87.5					
88.5				Turonian	
91				Cenomanian	(5)
97.5			Lower	Albian	(4) (D)
113				Aptian	(C)
119				Barremian	(B)
125				Neocomian: Hauterivian	(A) (3)
131				Valanginian	
138				Berriasian	
144		JURASSIC	Late	Tithonian	
150				Kimmeridgian	
156				Oxfordian	
163			Mid	Callovian	(2)
169				Bathonian	
175				Bajocian	
181				Aalencian	
188			Early	Toarcian	
194				Pliensbachian	
200				Sinemurian	(1)
206					

Q = Quaternary
Co = Coniacian
* Last 10^4 years of Quaternary is the Holocene

1. Asian fragments begin to separate from Australian Gondwanaland
2. These fragments completely detached
3. Rifted continental fragments above sea level, and now isolated within Tethys Ocean
4. The fragments above sea, available from now on as stepping-stones between Laurasia and Gondwanaland
5. New Guinea–Australia begins to separate from Antarctica
6. Northwards movement of Australia–New Guinea reactivated (continues till today)
7. Collision between Laurasia and Gondwana in central Malesia

A First recognizable angiosperm pollen type appears (monosulcate; *Clavatipollenites*)
B Second angiosperm pollen type (tricolpate) found in Brazil and west Africa
C First record from Australia of *Clavatipollenites* pollen type
D First records from Australia and north temperate latitudes of tricolpate pollen.

The living mammals (Musser, Chapter 7), flowering plants (van Balgooy, Chapter 8), and Lepidoptera (Holloway, Chapter 9) of Sulawesi all show, on close analysis, stronger connections to the north (Philippines) and south (Lesser Sunda Islands) than to Borneo and the rest of the Sunda Shelf west across the Makassar Strait. Wallace's line is a very real boundary. For all these groups Sulawesi is a unit; present ranges in no way reflect its possible dual origin by collision of Laurasia and Gondwana. Fossil vertebrates, including giant tortoises and proboscideans, have been found only in the southwest arm and this could indicate that region's origin as a separate island.

The Malay archipelago is today a geographically and geologically highly complex region of islands lying between two continents. It has a highly complex origin from the Mesozoic onwards. Plants and animals have tracked the evolving geography, themselves evolving, and have also been influenced by cyclic changes in climate. George, in the final essay (Chapter 10), draws the many threads together. Problems remain (the Fagaceae still prominent amongst them). We can point to the need for further geological work, for example on dating the rifting events, which is of great significance for botanists. We badly need more fossil evidence for early angiosperms, especially from northern Australia and the Gondwanic terranes embedded in south-east Asia. The biogeographical evolution of the Malay archipelago still has mysteries, Wallace's line still has us in its thrall.

2 DISPERSAL OF GONDWANALAND: RELEVANCE TO EVOLUTION OF THE ANGIOSPERMS

M. G. Audley-Charles

Two aspects of Gondwanaland fragmentation and dispersal are discussed. Firstly, the geological indications that Burma, Thailand, Malaya, and Sumatra comprise continental fragments rifted from northern Australia–New Guinea in the Mesozoic are reviewed. The possibility that these regions, which contain fossil land plants, were above sea level in the late Jurassic (160 Ma) and early Cretaceous (130 Ma) and acted as Noah's arks for the ancestral angiosperms is discussed. The indications that large parts of these regions remained above sea level throughout the Cretaceous and Cainozoic, thus providing opportunities for land plant colonization, are considered.

An outstanding problem concerns the precise date of separation of these Asian continental fragments from the Australia–New Guinea part of eastern Gondwanaland. There is some geological evidence for the separation being as late as late Jurassic Oxfordian stage (160 Ma), but some indications have been taken to mean that the rifting occurred as early as Permian. The conclusion reached here is that the weight of evidence is strongly in favour of Burma, western Thailand, Malaya, and Sumatra having been rifted from the north Australia–New Guinea continental margin during the Jurassic. These south-east Asian blocks became relatively isolated within the Tethys Ocean between Gondwanaland and the Asian mainland for possibly as long as 60 Ma, between about 160 Ma and 100 Ma. Equally important are the indications that from the late Cretaceous (100 Ma) onwards they provided an archipelago of islands between the Asian mainland and Australia–New Guinea which could have permitted land plant dispersal to occur in both directions throughout that time.

Secondly, the slow rifting of Australia–New Guinea from Antarctica at about 90 Ma, with a subsequent phase of rapid drift northwards from 44 Ma, until Australia–New Guinea collided with the south-east Asian volcanic island arcs between 15 Ma and 3 Ma is discussed. The collisional events involving northern New Guinea, Sulawesi and the Banda Arc resulted in the uplift of new land areas in Malesia. The Cainozoic convergence of Australian Gondwanaland on Asia has given rise to new islands by volcanic activity and collision, thus the opportunities for land plant dispersal between Asia and Australia have been substantially increased since the first collisions about 15 Ma.

INTRODUCTION

In the late Jurassic all the southern continents (South America, Africa, India, Australia, and Antarctica) were joined (Fig. 2.1) to form the supercontinent Gondwanaland. By the end of the Cretaceous these continents had rifted apart and India had already drifted about 1200 km northwards, being then separated from southward-drifting Antarctica by about 4000 km of new ocean crust; Africa had separated from Antarctica by about 2800 km, South America was still in contact with Antarctica, and Australia may have separated by about 200 km from Antarctica (Smith *et al.* 1981; Cande and Mutter (1982). The evidence revealing the pathways and timing of this dispersal ('flight from the pole') of the southern continents is preserved in the palaeomagnetic 'stripe' anomalies in the crust of the ocean floor that formed between the separating continents.

It has generally been thought that the early angiosperms evolved in western Gondwanaland

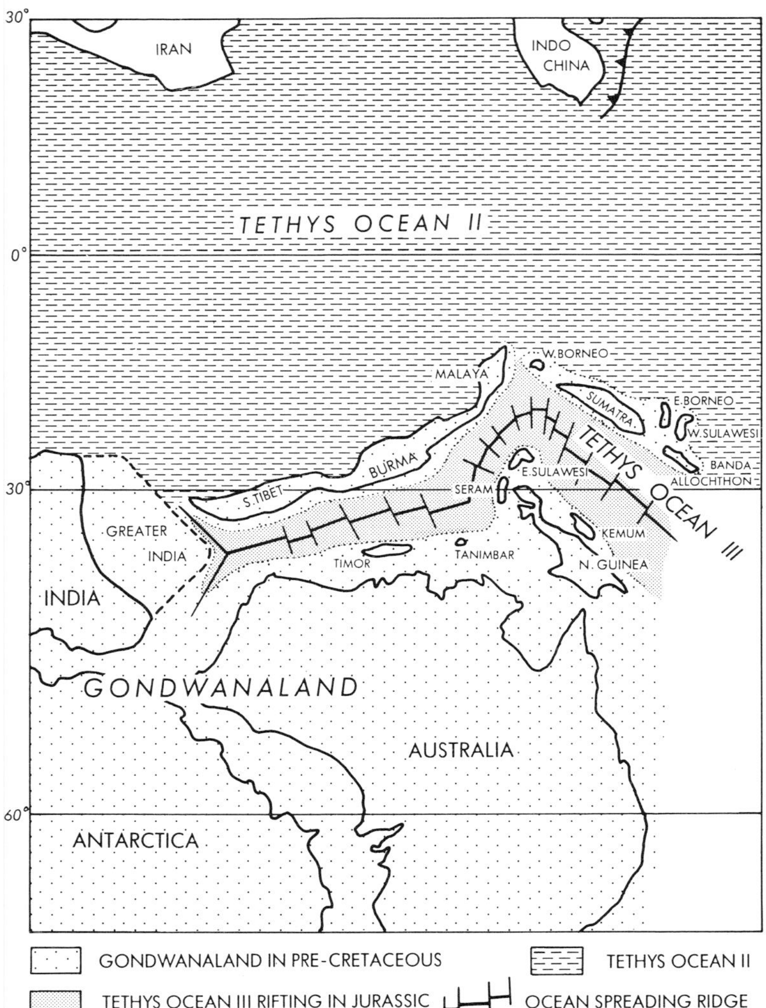

Fig. 2.1. Gondwanaland reconstruction for 160 Ma late Jurassic (after Barron *et al.* 1981; Smith *et al.* 1981). South-east Asian and South Tibet continental fragments just rifted from eastern Gondwanaland (after Audley-Charles 1983, 1984). Land plant dispersal between mainland Asia and Australia unlikely. Present coastlines for reference only.

(i.e. in South America and Africa) during the late Mesozoic. The oldest fossils identified as angiosperms are of Barremian and Aptian age (120 Ma) according to Takhtajan (this volume, Chapter 3). However, these are said to show such well-developed features that they imply the angiosperms must have originated much earlier, at least as far back as the late Jurassic. The apparently barren pre-Cretaceous angiosperm fossil record has been explained by the 'upland origin' hypothesis which postulates that these plants originated on mountainous slopes where suitable conditions for fossilization are rare. The sensitivity of land plants to climate may be a contributory factor to the apparent diachronism of late Cretaceous and early Tertiary plants in Arctic and mid-latitude North America (Hickey *et al.* 1983).

The presence in south-east Asia of living 'primitive' angiosperm families such as Magnoliaceae, Winteraceae, Proteaceae, and Fagaceae, together with the great richness and diversity of the south-east Asian flora, has led some botanists to regard south-east Asia and the adjacent south-west Pacific region as the cradle of the angiosperms, meaning that this was the main centre of evolution from which these plants radiated to colonize the rest of the world. This argument was put forward as long ago as 1969 by Takhtajan, and is reconsidered and restated in the light of new evidence by him in Chapter 3 of this volume. This concept was developed before that of sea-floor spreading. The application of plate tectonics to south-east Asia and the south-west Pacific region has demonstrated that the present configuration of islands and small ocean basins between Australia and the south-east Asian mainland is the consequence of Australia having converged about 300 km on Asia during the last 44 Ma. As Whitmore (1981) pointed out, the present pattern of islands joining Australia to the continent of Asia came into existence during the last 15 Ma which is too short a time to account for the present global range of flowering plants to have evolved and radiated from these tropical islands. There is also evidence which suggests that some plant groups may have entered the Malay archipelago from both the north-west and the south-east. Some of these still have bicentric distributions within the region (Whitmore 1981). In 1981 it appeared that various groups of palms, which are a big tropical family strongly represented in Malesia, might be amongst those which had entered both from Laurasia and Gondwanaland (Dransfield 1981, especially Figs 6.7, 6.9, 6.10). Further taxonomic study of palms reiterates that the family have Gondwanic and Laurasian genera and, as a family, have entered Malesia from both ends. There has been intermingling of the two elements which might be more ancient than the mid-Miocene 15 Ma collision which created the present-day archipelago (Dransfield, this volume, Chapter 6). The fossil pollen record in Australia indicates land connections between Australia and Asia at least as far back as the early Tertiary (Quilty 1984), and Truswell *et al.* (this volume, Chapter 4) also suggest that such connections may have existed in the mid-Cretaceous.

At the same time as botanists have been discussing the existence of land connections between Asia and Australia during the late Mesozoic and early Tertiary in order to account for both the present-day bicentric distributions of certain floral elements in Malesia and the fossil evidence for migration of plants into Australia from the north from mid-Cretaceous times onwards, geologists have been trying to identify the continental fragments that were rifted from the northern margin of Australia–New Guinea in the Mesozoic. Ridd (1971) and Tarling (1972) suggested that south-east Asia formed part of Gondwanaland during the Palaeozoic. Hamilton (1979) suggested that Indo-China, the Malay peninsula and Sumatra were adjacent to what is now central New Guinea until they were rifted from there in the Jurassic. Northern New Guinea did not become attached to central New Guinea until mid-Tertiary times. Mitchell (1981) proposed that Burma and the Thai–Malay peninsula lay adjacent to northern Australia during the Palaeozoic. Stauffer and Gobbett (1972) and Stauffer (1974) argued that the faunas and lithofacies of the late Palaeozoic rocks of these parts of south-east Asia indicate that they were deposited in the tropics or subtropics and hence that south-east Asia could not have formed part of Gondwanaland at that time when it was so far south. Audley-Charles (1983) following the suggestions of Hamilton and Mitchell, and taking account of the sparse palaeomagnetic data for the region, showed that the objections of Stauffer and Gobbett were invalidated by plotting the position of the south-east Asian continental fragments on the northern margin of Australia–New Guinea according to the global reconstructions of Smith *et al.* (1981). Later Audley-Charles (1984) plotted on his reconstruction of eastern Gondwanaland the position of the late Carboniferous glacial deposits reported from various parts of south-east Asia and from South Tibet. This strengthened the argument for south-east Asia and South Tibet having been attached to northern Australia–New Guinea during the late Palaeozoic (Fig. 2.2).

The first part of this paper deals with the evidence concerning the timing of the rifting and

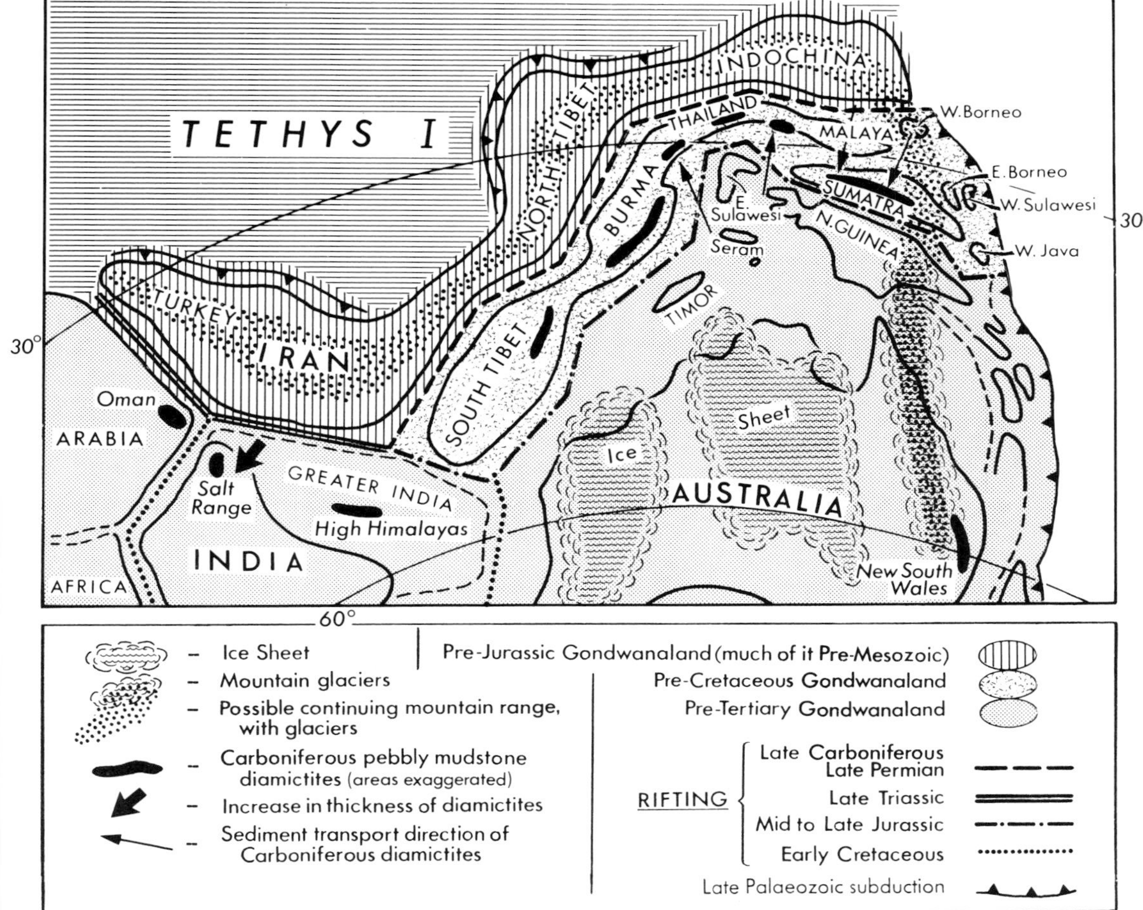

Fig. 2.2. Reconstruction of eastern Gondwanaland to show the location of late Carboniferous glacial diamictites in south-east Asia, Tibet, Himalayas, and Oman (after Audley-Charles 1984). Distribution of Carboniferous mountain glaciers (after Veevers 1984). Speculative continuation of mountain belt northwards to explain directions of sediment transport in Sumatran diamictites. Present coastlines for reference only.

drifting of the continental fragments that were removed from the northern margin of Australia–New Guinea during the Mesozoic to become part of Asia. Like the botanical evidence it is open to different interpretations. Palaeogeographical reconstructions of the region between the Asian and Australian continents during the late Mesozoic and early Tertiary can provide the land areas to explain the distribution of flowering plants if the ancestral angiosperms had evolved by the Oxfordian stage of the Jurassic (160 Ma) and were present in eastern as well as western Gondwanaland at that time.

The second part of the paper deals with the geological evolution of the Malesian region during the last 30 Ma. Recent information bearing on the age of the Banda Sea, the geological evolution of Sulawesi and the Moluccas as well as the mode of evolution of the Banda volcanic (inner) arc is reviewed. All these matters have implications for the availability of land that could be colonized by plants in this region of Cainozoic convergence and Miocene–Pliocene collision between Laurasia and Gondwanaland.

JURASSIC RIFTING OF THE NORTHERN AUSTRALIA–NEW GUINEA CONTINENTAL MARGIN

In Chapters 3 and 4 of *Wallace's line and plate tectonics* (Whitmore 1981) the view was taken that throughout the greater part of geological history the two halves of the Malay archipelago remained far apart; Malaya, Borneo, and Sumatra being part of Asia while eastern Indonesia (Timor, Tanimbar, Seram, Buru, Buton) was part of Gondwanaland separated from Asia by the wide Tethys Ocean. That view has been greatly modified (Fig. 2.2) largely as a result of palaeontological discoveries in New Guinea (Archbold *et al.* 1982) and as a consequence of recognizing the significance of the distribution of Carboniferous marine glacial deposits in southeast Asia (Cameron *et al.* 1980). The arguments of Buffetaut (1981) and Buffetaut and Ingavat (1980, 1982) that the Mesozoic vertebrates they have described from Laos and Thailand are evidence for land connection of eastern Thailand (Korat Plateau) and Laos with Laurasia in the pre-Cretaceous Mesozoic do not contradict, as they seem to think, the proposal that western Thailand, Burma and Malaya did not finally separate from Gondwanaland until the Jurassic. Despite extensive exposure of similar continental facies in western Thailand, Malaya and Burma, these vertebrate fossils, typical of the Indo-China block, have not been reported from the western south-east Asia block (Fig. 2.3). As Mitchell (in press) and others have pointed out, these two parts of south-east Asia belonged in the Palaeozoic to two separate geological provinces, viz. the western south-east Asia block and the Indo-China block (Fig. 2.3). It was suggested by Audley-Charles (1983) that the Indo-China block of eastern Thailand, Laos, Cambodia and Viet Nam, separated earlier from Gondwanaland in the Permo-Carboniferous, whereas the western south-east Asia block did not separate until the Jurassic. Palaeomagnetic data (Achache and Courtillot 1985) indicate the presence of the Indo-China block in northern tropical latitudes close to Eurasia in the Triassic.

South-east Asia as part of Gondwanaland

There are now sufficient geological indications to allow us to consider with a high level of confidence that South Tibet, Burma, Thai–Malay peninsula and Sumatra formed part of eastern Gondwanaland attached to northern Australia–New Guinea during the late Palaeozoic (McTavish 1975; Hamilton 1979; Bally *et al.* 1980; Cameron *et al.* 1980; Mitchell 1981; Archbold *et al.* 1982; Audley-Charles 1983, 1984). What is less certain is precisely where these parts of Asia were located on the Australia–New Guinea margin and when they rifted from Gondwana. Mitchell (in press) for example, considers the rifting was Permo-Triassic, Parker and Gealey (1985), who interpret Burma as attached to north-west Australia until the mid-Cretaceous, share Mitchell's view that Sumatra and the Thai–Malay peninsula rifted from Gondwanaland during the Permo-Triassic but differ from Mitchell in regarding Sumatra and Thai–Malaya peninsula as having been a microcontinent located off shore northern Australia. Allegre *et al.* (1984) argued that if South Tibet was part of Gondwanaland it must have separated from Gondwana no later than the early Jurassic or late Triassic because they suggest it accreted to Eurasia before the end of the Jurassic. The geological evidence for these interpretations is qualitative; for example, the Permo-Triassic rifting of Burma and the Thai–Malay peninsula from northern Australia favoured by Mitchell is based on his interpretation of the Triassic Carnian siliciclastic sediments of western Burma and the Himalayas as flysch. A. H. G. Mitchell (personal communication, 1984) explained his interpretation in the following terms: 'If western south-east Asia were part of Gondwanaland in the Triassic, the flysch must have accumulated in a kind of foreland basin on the continent'. However, very similar Triassic siliciclastic turbidites are well

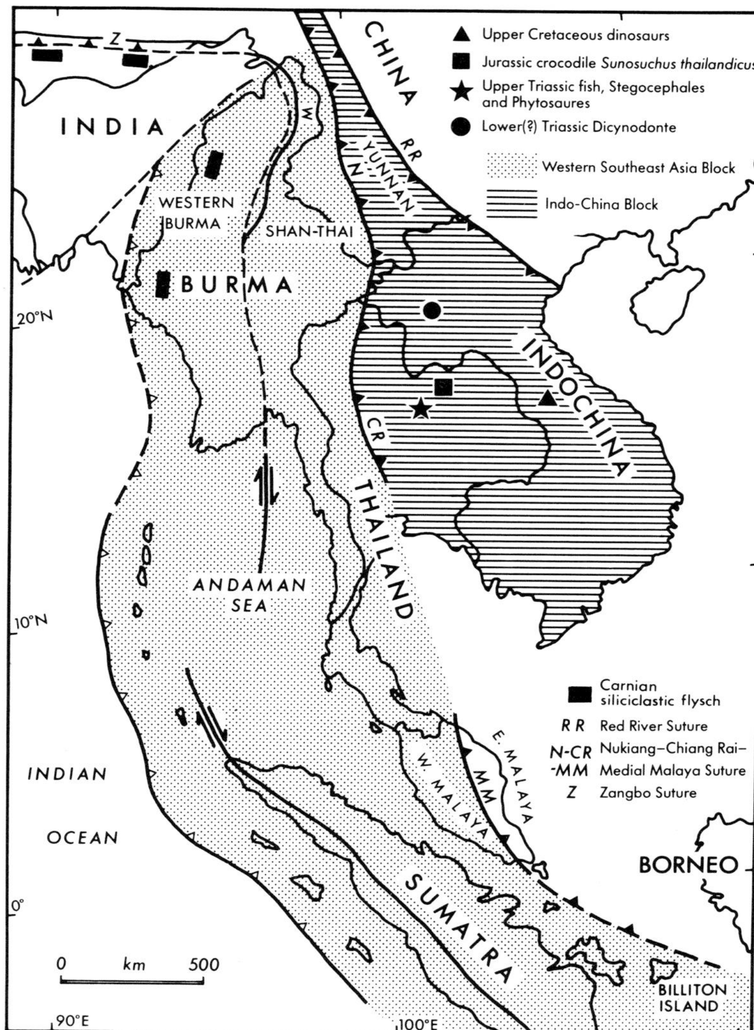

Fig. 2.3. Present outline of south-east Asia to show location of Mesozoic vertebrate fossils of Laurasian affinity (after Buffetaut 1981) and the principal tectonic sutures (after Mitchell, in press). Note the Laurasian vertebrates are confined to the Indo-China block which was part of Laurasia in the early Mesozoic (Achache and Courtillot 1985). The Western South-east Asia block (including the Shan-Thai block) is regarded here as having been part of Gondwanaland until the late Jurassic (see Figs 2.1, 2.4–2.9). The collision of this block with the Indo-China block generally regarded as a Triassic event is interpreted here as an early Cretaceous event.

exposed in the para-autochthonous Australian facies of Timor and Seram, where they appear to be an intracratonic foreland basin deposit (Audley-Charles 1983). Although that overcomes Mitchell's objection it does not of itself reinforce this interpretation. The strongest argument put forward by Allegre *et al.* (1984) in favour of their view, that South Tibet accreted to Eurasia before the end of the Jurassic, is based on the late Jurassic–early Cretaceous age of subaerial to shallow marine sediments that overlie the Dongqiao ophiolite, that is found thrust southwards over middle to upper Jurassic flysch (Girardeau *et al.* 1984). They correlate this event with that producing similar structures in Burma (Mitchell 1981) as part of a neo-Cimmerian (100–80 Ma) tectonic crisis along the south Eurasian margin. However, if these deposits, unconformable on the Dongqiao ophiolite, are early Cretaceous in age there is no conflict, as the postulated rifting of South Tibet could have moved the 3500 km northwards between 160 Ma and 120 Ma (at

8.75 cm/yr). Alternatively, these tectonic events might have occurred at the Gondwana margin and their effects carried later northwards to Eurasia as allochthonous terranes. The palaeomagnetic data (Allegre *et al.* 1984) suggests a late Cretaceous (80–60 Ma) palaeolatitude of between 10°N and 15°N for South Tibet. Thus, if South Tibet were rifted from north-west Australia (30°S) in the Oxfordian (160 Ma) it could have reached its late Cretaceous (80 Ma) palaeolatitude by ocean spreading at 5 cm/yr for 80 Ma. That may be compared with India following a similar path at 10 cm/yr for the first 30 Ma then 5 cm/yr for the next 40 Ma (Molnar and Tapponnier 1975).

Dating the rifting

Johnson and Veevers (1984) and Pigram and Panggabean (1984) have described the break-up unconformity through New Guinea and the northern Australia shelf where it ranges in age from Pliensbachian–Sinemurian (200 Ma) in Papua New Guinea to Bajocian (170 Ma) in Irian Jaya and Oxfordian (160 Ma) off shore Canning Basin (Fig. 2.4). Earlier Powell (1976) provided evidence from oil company seismic surveying and drilling in the northern Australian shelf for this break-up unconformity that can be accounted for only by continental fragments having been rifted from what is now the continental margin of

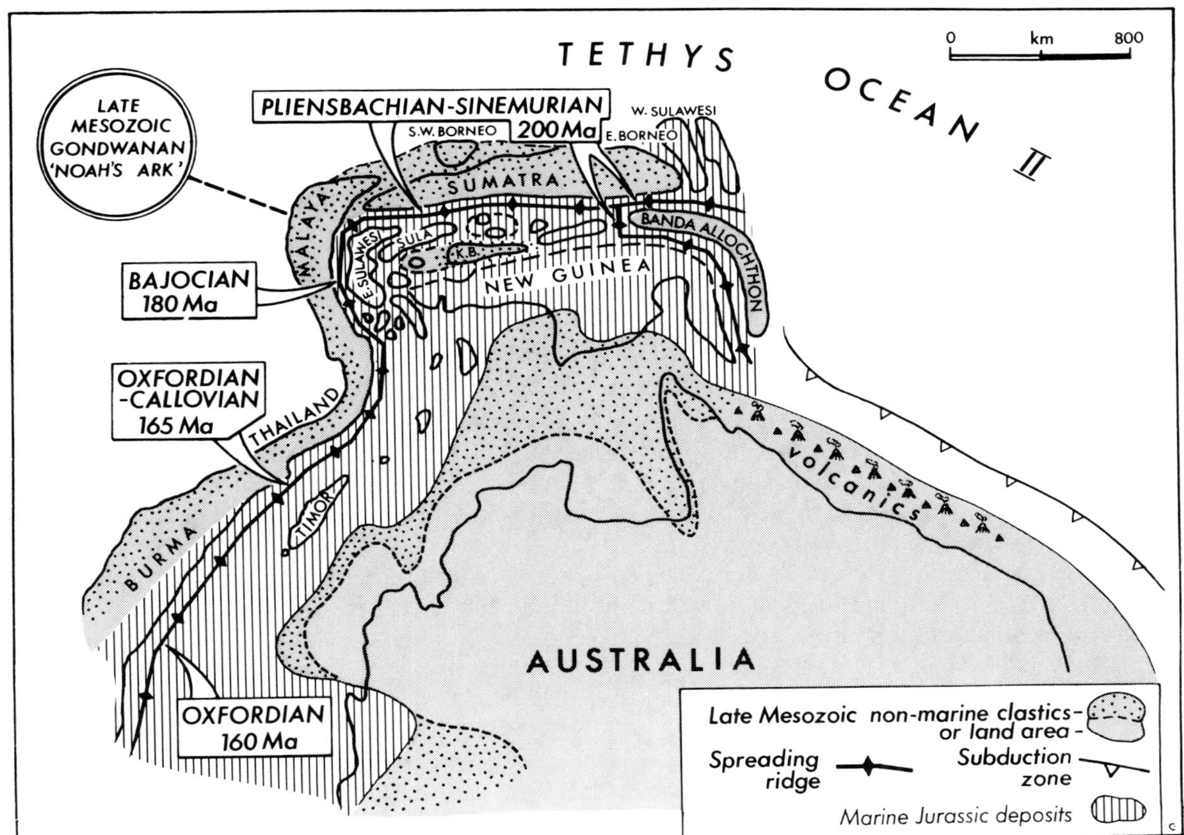

Fig. 2.4. Reconstruction of continental rift zone at the margin of eastern Gondwanaland during 200 Ma to 160 Ma (early to late Jurassic) modified after Pigram and Panggabean (1984). Location of south-east Asian fragments schematic. Outlines for reference only.

Australia–New Guinea. The collision of this rifted continental margin, in what is now central New Guinea, with a volcanic island arc complex during the middle Tertiary and the collision of the northern Australia rifted continental margin with the Banda volcanic arc destroyed the oceanic crust with its evidence of rifting contained in the palaeomagnetic 'stripe' anomalies. Thus the evidence for dating the rifting event has to be found in the stratigraphy and structures of the Australian continental margin rocks of the present Australian shelf and in the fold and thrust belt of the Outer Banda Arc islands of Timor, Seram, Buru, Buton and eastern Sulawesi. Detailed studies of the stratigraphy and structure in Timor indicate that the suggestion of Parker and Gealey (1985), for a rifting episode having removed Timor, Seram and other parts of the Outer Banda Arc region from the continental margin of Australia during the Mesozoic, is not only without support from the Mesozoic rocks exposed in those islands but their stratigraphy refutes such a suggestion. That part of the Earth's crust now represented by the Australian facies exposed in those islands has remained part of the Australian continent as demonstrated by the fossiliferous successions from Permian to Pliocene summarized in Audley-Charles *et al.* (1979). Detailed evidence is present in the stratigraphical successions exposed in the Kolbano, Betano, Iliomar and Aliambata regions of Timor (Barber *et al.* 1977). These sections and those of Nief in Seram are incompatible with these regions having been separated from the Australian continent by any rifting episode. What these exposures do indicate are considerable changes in depth of sediment accumulation, perhaps associated with the rifting episode in the middle Jurassic, but there is no possibility that they were detached from the Australian continent.

Although we can be confident that Burma, Thai–Malay peninsula (excluding eastern Thailand) and Sumatra were part of eastern Gondwanaland and were rifted from the continental margin of northern Australia–New Guinea during the Jurassic, the details of the palaeogeographical reconstruction remain a matter of speculation. The interpretation of South Tibet is more difficult. There are several geological features which imply that South Tibet formed part of the same geological province as Burma, and the Thai–Malay peninsula during the Palaeozoic and Mesozoic (Bally *et al.* 1980; Mitchell 1981; Audley-Charles 1983). However, the apparent absence in South Tibet of the deformation of the Triassic–Jurassic Indosinian orogeny involving westward and south-westward directed thrusting (Mitchell, in press) would seem to support the view that South Tibet separated from Gondwanaland in the late Palaeozoic or early Mesozoic, unless the overthrusting of the Dongqiao ophiolites in South Tibet represent a younger part of this Indosinian event.

Let us turn now to consider these Asian continental fragments drifting northwards from Australia–New Guinea towards Asia, between their separation from Gondwanaland which began in early Jurassic Sinemurian–Pliensbachian times (200 Ma) and was completed by late Jurassic Oxfordian times (160 Ma). From the point of view of their potential as Noah's arks for the early angiosperms, we need to seek three kinds of information: (a) were any parts of these Asian fragments above sea level during this time? (b) what was the effect of the late Cretaceous rise in sea level on the land areas in the region under consideration? (c) what constraints exist for reconstructing the movements of these Asian fragments (Figs 2.5–2.9) after they separated from Australia–New Guinea in the Jurassic and before they collided with Asia?

Exposed land in south-east Asia during the Jurassic and Cretaceous

One of the stratigraphical features that South Tibet (Bally *et al.* 1980; Achache and Courtillot 1984), Burma (Wolfart *et al.* 1984) and Thai–Malay peninsula (Gobbett and Hutchison 1973; Helmcke 1982; Hahn and Siebenhuner 1982) have in common is the widespread, non-marine,

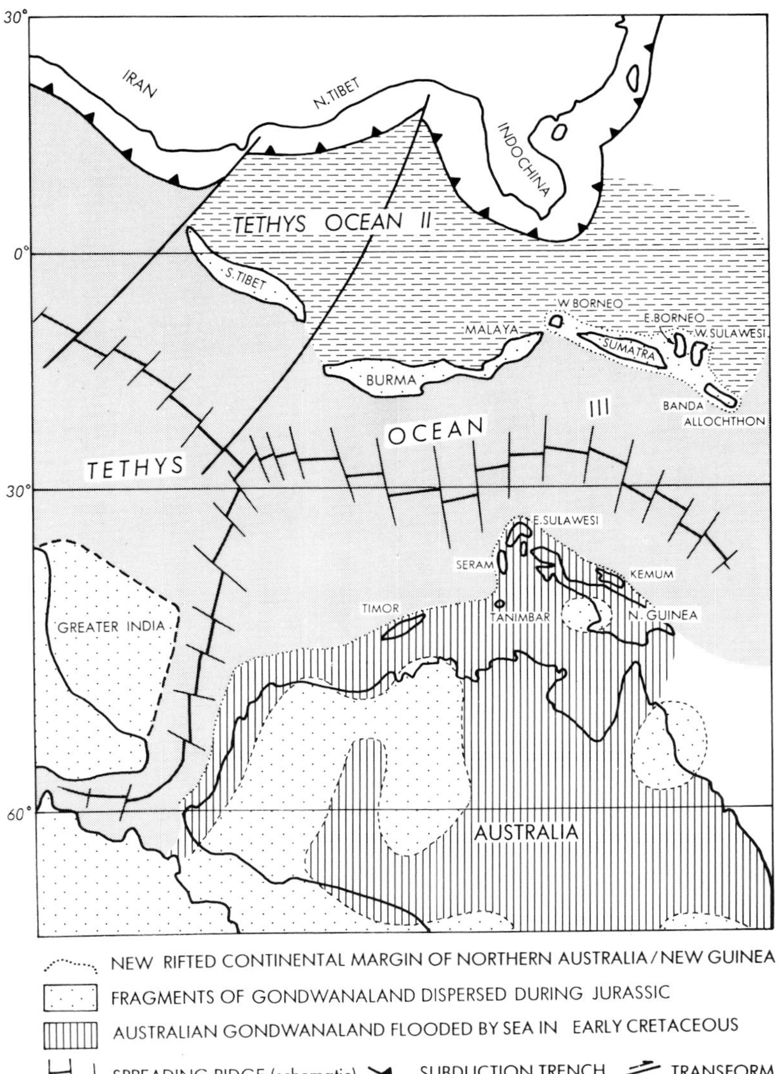

Fig. 2.5. Gondwanaland reconstruction for 120 Ma early Cretaceous (after Barron *et al.* 1981; Smith *et al.* 1981). Tibet and south-east Asia reconstruction follows Fig. 2.1. Spreading system and Eurasian subduction modified after Parker and Gealey (1985). Note the geographical isolation of south-east Asian Gondwanaland 'Noah's arks'. Land plant dispersal between mainland Asia and Australia–New Guinea unlikely. Present outlines for reference only.

often red-bed, siliciclastic facies of Jurassic and Cretaceous age. The lack of reports of this facies from Sumatra is surprising, but recently Cameron *et al.* (1980) have reported 'plant-bearing arenites in Riau which may correlate with the post-orogenic Tembeling Formation in Malaysia which is largely fluviatile'. Taken together these observations record evidence for widespread exposed land areas in South Tibet, Burma, Thailand, and the Malay peninsula. It is quite possible but at present uncertain that parts, even perhaps large parts, of Sumatra were also exposed as land during the Jurassic and Cretaceous. This suspicion is strengthened by the relatively few reports of any Jurassic and Cretaceous rocks in Sumatra (van Bemmelen 1949; Hamilton 1979).

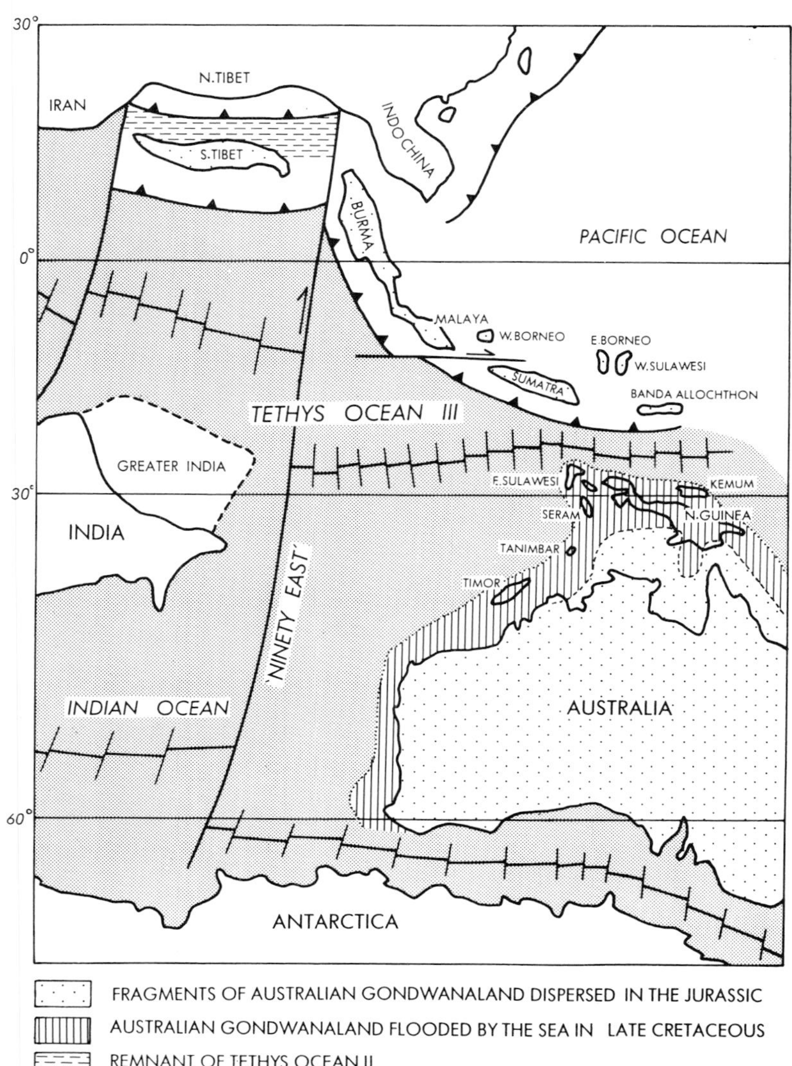

Fig. 2.6. Gondwanaland reconstruction for 90 Ma late Cretaceous (after Barron *et al.* 1981; Smith *et al.* 1981). South Tibet by anticlockwise rotation now at margin of North Tibet. Northward movement of Australia and adjacent Antarctica (from Barron *et al.* 1981; Smith *et al.* 1981). Note postulated clockwise rotation of south-east Asian fragments from Gondwanaland brings New Guinea close to Asian fragments. Land plant dispersal between mainland Asia and Australia possible. Present outlines for reference only.

EFFECTS OF CRETACEOUS SEA-LEVEL RISE

It is generally agreed that during the Cretaceous period there was a world-wide rise in sea level (Vail *et al.* 1977; Hallam 1984). What is less certain is to what extent this was achieved in steps, with intervals of still-stand or fall, whether the rises were synchronous in different areas; and of course there is debate over the origin of the rise. In our concern with the exposed land areas available for colonization by land plants in the region of Australia and those parts of south-east Asia that were attached to northern Australia–New Guinea during the Jurassic (viz. Burma, Thai–Malay peninsula, Sumatra, and possibly also South Tibet), the effect of this world-wide sea-level rise in the Cretaceous is relevant.

The Australian Bureau of Mineral Resources *Earth Science Atlas* (Wilford 1983) reveals in a

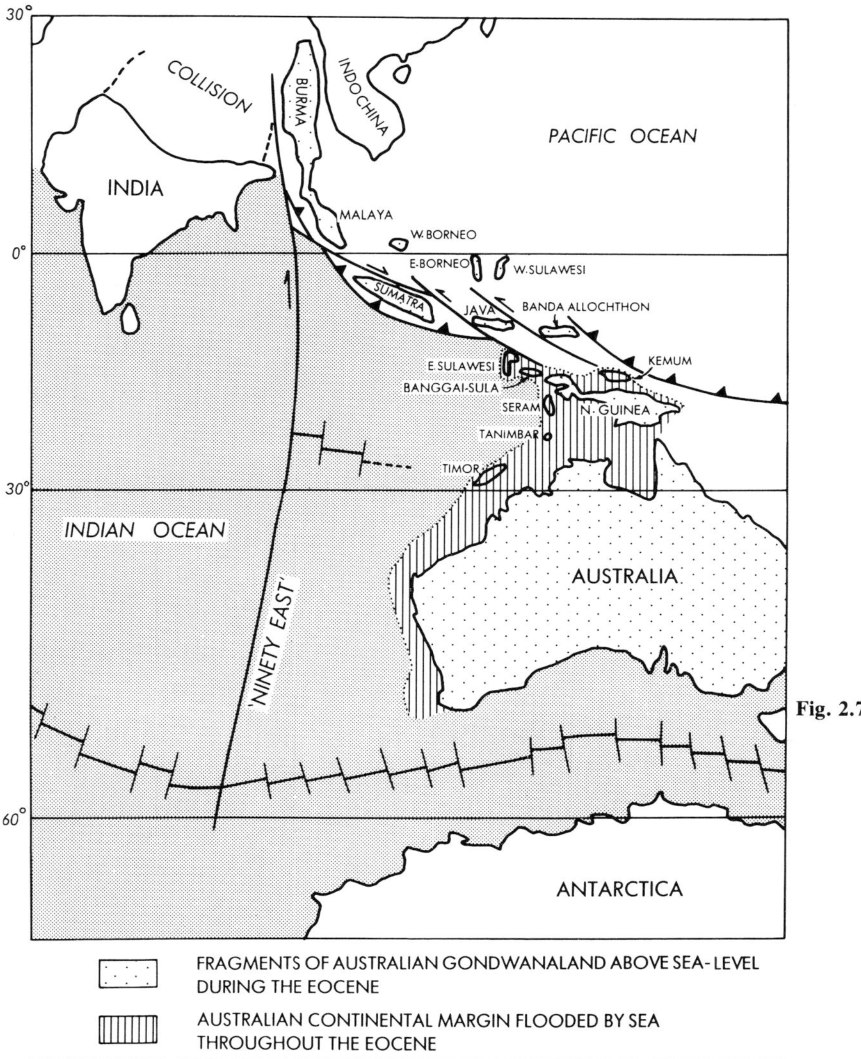

Fig. 2.7. Reconstruction of convergence zone between Australia–New Guinea and south-east Asia for 40 Ma late Eocene (after Barron *et al.* 1981; Smith *et al.* 1981). Relations between west and east Borneo uncertain. Note Australia's northward movement keeping it relatively close to south-east Asian fragments. Land plant dispersal between mainland Asia and Australia possible. Present outlines for reference only.

series of palaeogeographical maps that most of the present land area of Australia was land during the early Jurassic. The world-wide rise in sea level, beginning during the Jurassic and continuing throughout the Cretaceous, had its greatest impact on Australia during the early Cretaceous, when large parts of central Australia, including most of the Great Artesian Basin, were flooded by marine transgression. However, even this left large tracts of land exposed in northern, western and eastern Australia. During the late Cretaceous, when the world's continents were largely inundated by marine waters, almost all of Australia was above sea level. From this it can be seen that land plants in Australia should not have experienced any major catastrophe from changes in sea level during the Mesozoic.

A similarly anomalous picture seems to emerge from the reports of the Jurassic and Cretaceous strata of the south-east Asian blocks

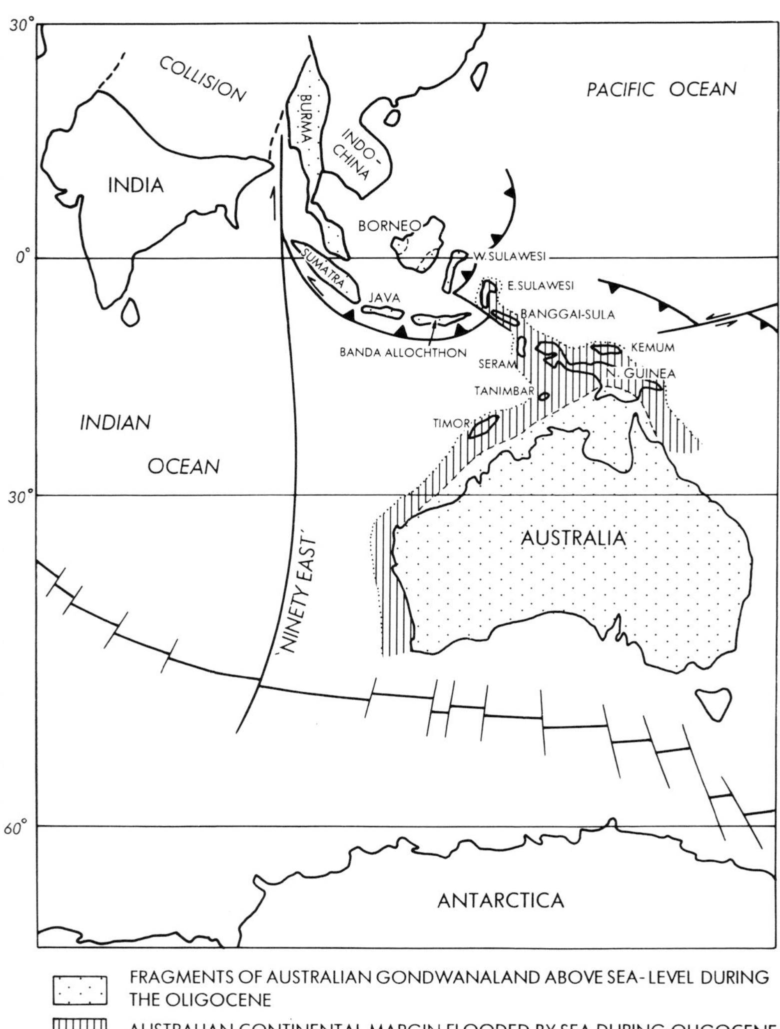

Fig. 2.8. Reconstruction of convergence zone between Australia–New Guinea and south-east Asia for 30 Ma Oligocene. Volcanism in Sunda Arc and southwest Pacific arcs. Central New Guinea collision with volcanic arc associated with convergence and postulated sinistral movement. Land plant dispersal between mainland Asia and Australia possible. Present outlines for reference only.

of Burma, Thailand, Malay peninsula and South Tibet. These regions also seem to have remained in part, or even large part, above sea level during the Jurassic and early Cretaceous, judging from the reported presence of non-marine deposits (referred to above). The scarcity of marine sediments of Jurassic and Cretaceous age reported from Sumatra suggests comparison with Australia and these other parts of south-east Asia. The history of Australia and south-east Asia during the early Tertiary is largely one of extensive land areas being exposed, but it is uncertain to what extent the late Cretaceous world-wide maximum sea-level rise resulted in Sumatra, Thailand, Malay peninsula, and Burma being invaded by the sea. The negative evidence that few deposits of such a sea have been found in these areas is all we have to go on, apart from the presence of non-marine deposits of early Cretaceous and early Tertiary age in Burma (Wolfart *et*

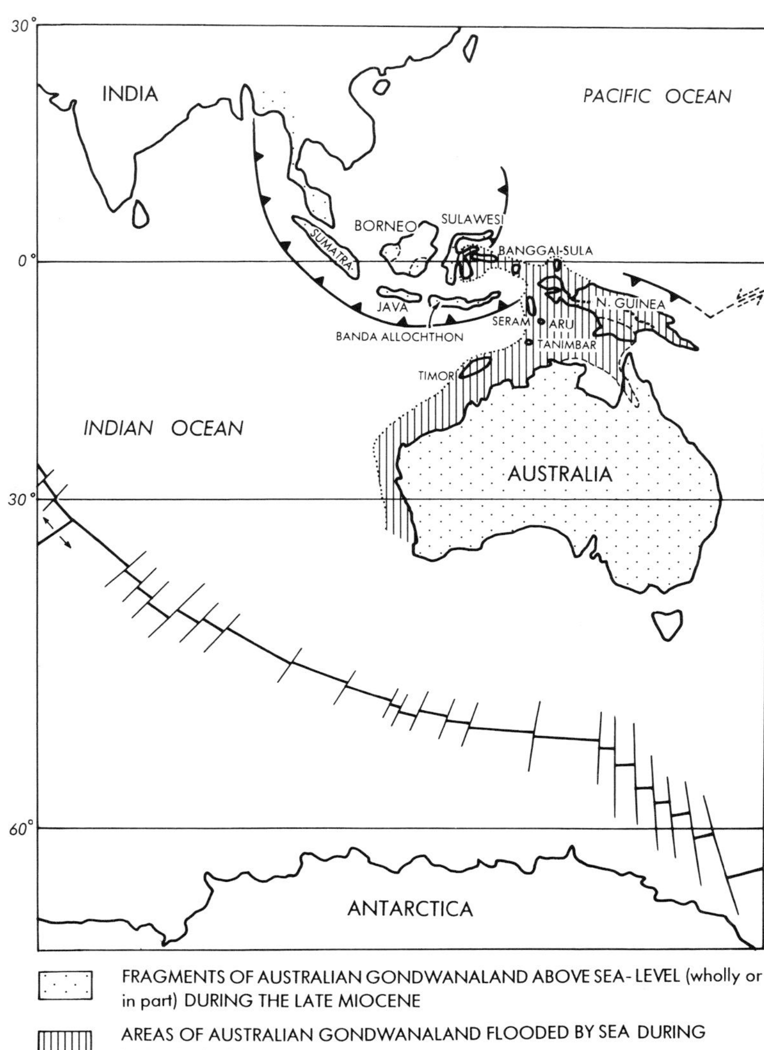

Fig. 2.9. Reconstruction of convergence and collision zones between Australia–New Guinea and south-east Asian fragments and volcanic arcs for 10 Ma late Miocene. East–west Sulawesi collision was 15 Ma middle Miocene. Seram–Banda volcanic arc collision was in late Miocene to early Pliocene at about 5 Ma. Timor–Banda volcanic arc collision was in mid-Pliocene about 3 Ma. Land areas between mainland Asia and Australia increased after mid-Miocene times as a result of volcanic and collision processes. Outlines are for reference only.

al. 1984), Thailand (Hahn and Siebenhuner 1982), and the Malay peninsula (Gobbett and Hutchison 1973). According to Cameron *et al.* (1980) 'there is no sedimentary record on mainland northern Sumatra for the period between the late Cretaceous orogeny and the Eocene. Sumatra was probably emergent during this time'. During the Tertiary all these parts of southeast Asia had considerable tracts of land above sea level with fluviatile, deltaic and lacustrine deposits being preserved.

LAND AREAS BETWEEN AUSTRALIA AND ASIA DURING LATE MESOZOIC AND EARLY TERTIARY

It is widely held that the flowering plants evolved in west Gondwanaland and radiated into east

Gondwanaland (Australia) and via South America and Africa into Laurasia. Hickey *et al.* (1983) suggested that the Arctic may have been an important centre for flowering plant evolution during the late Cretaceous. Dransfield (1981, and this volume, Chapter 6) and Whitmore (1981) pointed out the bicentric distribution of some important floristic elements in Malesia that have already been referred to in the Introduction to this chapter. Both authors discussing palms and other plant groups have suggested this distribution indicates that these plants entered Malesia from both east Gondwana (Australia–New Guinea) and from west Gondwana via Laurasia. Some of these patterns seem incompatible with a post-Miocene (i.e. post-collision of Australia with the Asian island arcs) migration.

According to Quilty (1984) and Truswell *et al.* (this volume, Chapter 4) the record of Australian fossil pollen indicates that land connections between Australia and Asia existed as far back as the mid-Cretaceous. Christophel and Basinger (1982) reported fossil flowers belonging to *Diospyros* from the Eocene of Australia, which according to D. C. Christophel (personal communication, 1985) 'is more similar to certain species now growing in Malaysia and the Philippines than it is to the extant Australian species'.

The following lines of evidence, discussed above, all point to the presence between Australia and mainland Asia during the Cretaceous and early Tertiary of a continental fragment or number of separate fragments rifted from the continental margin of northern Australia–New Guinea during the Jurassic.

(a) The oil company seismic reflection and drilling data from the north Australian shelf indicate there was a major phase of rifting that involved the separation of a continental mass to the north and formation of new oceanic crust of which the Argo Abyssal Plain in the northeast Indian Ocean is part.

(b) Geological field mapping has shown that this rifting phase started in the Sinemurian–Pliensbachian (200 Ma) near the Kubor Range of Papua New Guinea, migrated anticlockwise with rifting in Irian Jaya in the Bajocian (170 Ma) according to Pigram and Panggabean (1984). It reached the region north of Timor in Callovian–Oxfordian times (165 Ma) and (according to oil company data from the shelf) the region of the Argo Abyssal Plain and Rowley Shoals opposite the Canning Basin in the Oxfordian (160 Ma).

(c) Sundaland, comprising Burma, Thai–Malay peninsula, and Sumatra has been identified as the principal continental block that was rifted from northern Australia–New Guinea during the Jurassic with the possibility that South Tibet was the westernmost part of this block to be rifted in the Jurassic, although Allegre *et al.* (1984) have put forward arguments in favour of South Tibet being rifted from Gondwana much earlier, in the Mesozoic or late Palaeozoic. There is strong evidence that South Tibet was already joined to North Tibet as part of Eurasia by late Cretaceous (80 Ma). Powell and Johnson (1980) used palaeomagnetic data to plot the migration track of India after it broke from Gondwanaland in the Cretaceous and showed that India's ocean-floor path took it across the position now occupied by western Malaysia and Sumatra. They showed that the northern part of Sundaland must have rotated at least 550 km westward with respect to India during the last 50 Ma (since the early Eocene) with most of the movement occurring since the end of the Oligocene (25 Ma). This limited information has been used in this chapter to guide the palaeogeographical reconstruction of the movement of Sundaland after it rifted from Australia–New Guinea in the Jurassic. Apart from the magnetic 'stripe' anomalies of the Indian Ocean floor and those less certain anomalies of the floor of the Banda Sea (Lapouille *et al.*, in press) there are very few palaeomagnetic data to constrain the plot of the migration path of Sundaland from Gondwana to its present location, which it must have achieved not later than 15 Ma (middle Miocene) when the collision with Australia, following closely in its wake, occurred.

(d) The evidence from the fossil pollen and plants in Australia and the present distribution of plants in Malesia indicate that there was a land area located between Asia and Australia from which Gondwanan flora were able to migrate in both directions during the early Tertiary.

(e) There is geological evidence to show that land areas potentially available for plant colonization were present during the Jurassic, Cretaceous, and early Tertiary in the continental blocks that were rifted from Australia–New Guinea in the Jurassic.

MOVEMENTS OF THE GONDWANA NOAH'S ARK

Burma, Thai–Malay peninsula, Sumatra, and possibly South Tibet have the geological attributes, and could have occupied the appropriate position, to act as Noah's ark in carrying an evolving angiosperm flora from east Gondwanaland to Asia, and to have allowed floral migrations in the opposite direction into Australia during the late Mesozoic and early Tertiary.

The series of palaeogeographical maps (Figs 2.5–2.9) illustrate this potential although the precise movements of the rifted fragments that now constitute Sundaland and South Tibet are uncertain and hence open to some modification.

LATE CAINOZOIC GEOLOGICAL EVOLUTION OF EASTERN INDONESIA

In Chapters 3 and 4 of *Wallace's line and plate tectonics* (Whitmore 1981) it was noted that during the Cainozoic Australia–New Guinea moved northwards converging on south-east Asia. Four lines of evidence for this convergence were referred to. The first line of evidence is provided by the growth and spreading of the Antarctic Ridge from 44 Ma (which has been discussed more recently by Johnson and Veevers 1984). Secondly, palaeomagnetic data indicate a convergence between Asia and Australia during the Cainozoic and spreading of new ocean south of the northward-drifting Australia–New Guinea (Smith *et al.* 1981). Thirdly, the pattern of northward-inclined earthquakes between Australia and Indonesia (Hamilton 1974) and the associated line of volcanicity forming the Inner Banda Arc indicates subduction of lithosphere belonging to the Australia–New Guinea plate below the Banda Arc and its relation to the northward movement of Australia–New Guinea away from Antarctica. The fourth line of evidence is provided by the deformation of the crustal rocks in the Outer Banda Arc. This represents the crustal shortening in the zone of convergence.

Since 1981 new information has been published on the age and mode of formation of the Banda Sea, on the composition of the different geological provinces in Sulawesi, and on the geological affinities of the Moluccas and western Irian Jaya. New models for the geological evolution of the Banda Arc have also been developed. All of this has implications for the distribution of land and sea during and after the collision between Australia–New Guinea and the Asian islands, and hence for the potential dispersal paths of colonizing land plants radiating from Australia–New Guinea and from mainland Asia.

Recognition of the collisional boundary between Gondwanaland and Laurasia

The most recent review of the geology of Sulawesi by Sukamto and Simandjuntak (1983) recognized three geological provinces: a western province in the west and north arms, an eastern province in the east and south-east arms, and the Banggai–Sula province in the Banggai–Sula islands; the last is also present in the east arm. The western province is characterized by neritic siliciclastic and carbonate sediments and volcanics ranging in age from late Cretaceous to early Miocene. The eastern province is characterized by a bathyal marine facies of calcilutite and chert of Jurassic–Cretaceous age. The Jurassic section includes shale and lithic arenites with calcilutites and cherts and contains some molluscs. These Mesozoic rocks have been strongly deformed and are thought to be deposited on ultramafic basement but that is uncertain. The overlying Palaeogene neritic section, in which limestones play an important part, is also found in the Banggai–Sula province. This Banggai–Sula province is characterized by shallow marine late Triassic sediments that are overlain by fossiliferous Jurassic and Cretaceous neritic deposits in contrast with the deep marine facies of the eastern province. Similar neritic facies but with a higher carbonate content characterize the Palaeogene and early Miocene section of both the eastern and the Banggai–Sula provinces. All three provinces display similar middle Miocene and younger deposits in the form of post-orogenic detritus. In addition there is strong volcanic activity in the Neogene of the western province. One of the important features is the way that ophiolites associated with both blueschists and greenschists characterize the eastern province and separate the eastern from the western province. This emphasizes the tectonic suture between western and eastern Sulawesi. The relationship between eastern province and the Banggai–Sula province appears to be one of structural and stratigraphical transition that has been telescoped by tectonic imbrication and thrusting. The apparent absence of the Palaeogene and early Miocene deep marine facies from the eastern province may be attributed to imbrication and thrusting that has cut out that part of the section.

The western part of Sulawesi belongs to a very different geological domain from the eastern province and the Banggai–Sula province. The eastern areas show many lithofacies and palaeontological affinities with Gondwanaland by their close comparison with Australia–New Guinea. This work confirmed the mid-Miocene collision between the western and eastern provinces. The collision also resulted in the imbrication of the boundary of the eastern and Banggai–Sula provinces representing, as it does, the boundary between the deep water (i.e. continental slope and rise) deposits and the continental shelf deposits of the Banggai–Sula region. We can note that this same transition zone is exposed at Nief in northern Seram and in southern Timor at Iliomar, Aliambata, Betano and Kolbano where the same kind of imbrication with the same sense of vergence towards the Australia–New Guinea craton is displayed as it is in eastern Sulawesi and Banggai–Sula.

This raises the question of the palaeogeographical relationship between the western part of Sulawesi and Borneo. Hamilton (1979) argued that the similarity of the geological successions from the Cretaceous through the Tertiary on either side of the Makassar Strait suggested that until the middle Palaeogene time western Sulawesi was a part of eastern Borneo. He pointed out that the older sedimentary succession below the Mahakam delta may prograde westwards and hence imply an easterly source in Sulawesi. This and the presence of much normal faulting at the margins of the Makassar Strait suggest crustal stretching. Haile (1981) on the basis of some palaeomagnetic data concluded that south-west Sulawesi had been part of the same plate as the Malay peninsula and west Kalimantan since the late Mesozoic. Katili (1978) in a speculative paper suggested that Sulawesi originated quite independently from Borneo and that they collided in the late Pliocene closing the Makassar Strait. He postulated that the Makassar Strait evolved by the separation of Borneo and Sulawesi during the Quaternary. There does not seem to be any critical evidence to support this scheme. The presence of a thick pre-Miocene sedimentary succession below the Makassar Strait that has not been strongly compressed (Situmorang 1982) makes a strong case for regarding western Sulawesi as having been separated from Borneo by a flooded Makassar Strait at least as early as the early Miocene. Western Sulawesi and Borneo are both regarded (Audley-Charles 1983) as fragments of Gondwanaland (Figs 2.4–2.9) as is discussed later in this chapter (p. 23).

Most of the islands between New Guinea and Sulawesi appear to belong to Australian Gondwanaland. That includes Buton, Banggai–Sula, Buru, Seram, and the Lesser Sunda Islands of the non-volcanic Outer Banda Arc (Kai, Tanimbar, through to Timor and Roti). The evidence is based on the similarity of the lithofacies and faunas particularly of the Mesozoic and Permian strata (Audley-Charles 1978, Pigram and Panggabean 1984). Audley-Charles *et al.* (1979) presented data for extending this correlation through the Tertiary.

There is very little evidence available to demonstrate the geological affinity of the north Moluccas (Halmahera, Bacan, and Obi). Their present position with respect to Sulawesi, New Guinea, and the Banda Arc, as well as their position in relation to Benioff zones in the region and water depth all suggest that these north Moluccan islands are tectonically detached parts of New Guinea and hence their basement is part of Gondwanaland. On this basement some Neogene and Quaternary volcanoes have developed as a consequence of the plate movements. Pigram and Panggabean (1984) regard Obi and Bacan as a microcontinent derived from New Guinea.

The islands of Talaud, Mayu, and Tifore in the Molucca Sea appear from their position (Moore and Silver 1982), seismic reflection studies (Silver and Moore 1978), earthquakes (McCaffrey 1983), and from geological mapping on Talaud (Moore *et al.* 1981) to be imbricate zones of Tertiary and Quaternary sediments. They seem most easily understood in terms of their being part of the collision suture between Sundaland and Gondwana, the same collision suture as is exposed in central Sulawesi. A major difference may be that the Talaud–Tifore ridge may represent the fore-arc collision zone in an intra-oceanic setting, whereas central Sulawesi represents the continental margin of a rifted fragment of Gondwanaland that collided with the continental margin of Sundaland.

Origin and significance of the Banda Sea

The Banda Sea is the back-arc basin to the Banda Arc. The importance of the Banda Sea is that its age and mode of origin have a direct bearing on the interpretation of the collision of the Australian continent with the volcanic Inner Banda Arc which gave rise to the present configuration of islands immediately north of the Australian shelf.

Bowin *et al.* (1977) showed that the floor of the Banda Sea is oceanic-type crust. On the basis of the similarity in direction and the wavelength of the magnetic anomalies in the south Banda Sea they suggested that it might be Cretaceous in age. The presence in the islands of the Outer Banda Arc (Timor and Seram) of flat-lying overthrust sheets (Audley-Charles *et al.* 1979), that were interpreted on the basis of correlation as having been derived from the eastern margin of Sundaland, led Carter *et al.* (1976) to propose that the Banda Sea formed by back-arc spreading (Karig 1971) during the Neogene, a view shared by Hamilton (1977). Norvick (1979) also considered the allochthonous elements in the Banda Arc to have been derived from the margin of Sundaland but, taking note of the water depth and heat flow in the Banda Sea, concluded that it opened in the Palaeogene. Crostella (1977) and Katili (1978) considered the Banda Sea to be a piece of Pacific Ocean floor trapped by transcurrent faults and the subduction trench. Lapouille *et al.* (in press) argued that on the basis of water depth being as great as 5000 m, the heat flow values low (Jacobson *et al.* 1978), and the sedimentary layer thick (1–3 seconds two-way-time) which, using the relations between depth and heat flow versus age (Parsons and Sclater 1977), suggest that the floor of the Banda Sea dates back to at least the Palaeocene or Cretaceous. Lapouille *et al.* (in press) also used the magnetic anomaly pattern of the floor of the Banda Sea as further evidence for its being as old as early Cretaceous (130 Ma–120 Ma). They correlated these anomalies with those in the Argo Abyssal Plain of the north-east Indian Ocean and suggested that the floor of the Banda Sea is a piece of Tethys Ocean that became divided into the western Pacific and eastern Indian oceans by the northward-converging Australia–New Guinea continent. They envisaged this Tethys Ocean trapped during the Miocene by the present subduction system of the Timor–Seram troughs.

There is a strong case for the floor of the Banda Sea being early Cretaceous in age. However, the suggestion that the anomalies in the south Banda

Sea were continuous with those in the north-east Indian Ocean is difficult to accept on two grounds. The presence of the Banda Arc islands surrounding the Banda Sea on three sides, and the evidence in the islands of Timor and Seram of thrust sheets that moved southwards and northwards respectively away from the Banda Sea during the mid-Pliocene and early Pliocene, respectively indicate that the Banda Arc acquired its present sinuosity only after the overthrusting. Post-overthrusting sinuosity of the arc is difficult to accommodate geometrically with a floor of the Banda Sea so undisturbed that the orientation of its stripe anomalies has not been altered. Another observation that is difficult to fit to the idea of the Banda Sea floor anomalies being continuous with those in the north-east Indian Ocean is the presence in the middle of the south Banda Sea of the Banda ridges from which continental-type metamorphic and sedimentary rocks have been dredged (Silver *et al.* 1985).

The preferred interpretation adopted here is that the south Banda Sea floor is a piece of late Jurassic Tethys Ocean floor formed during the rifting of Asian continental blocks from Australian Gondwanaland. This differs from the Lapouille *et al.* (in press) model only in considering the rifting of the Australian continental margin of the Seram–Tanimbar region was late Jurassic (Fig. 2.4) rather than early Cretaceous as required by the Lapouille *et al.* model. The depth of the floor of the southern Banda Sea floor could have been modified by the Neogene–Quaternary tectonics and the volcanic events following collision. The Lucipara–Banda ridges in this model are regarded as another trailing and stretched part of eastern Sundaland rifted from Gondwanaland in the Jurassic. The dredging of Neocene vesicular andesites from these ridges (Silver *et al.* 1985) indicates the influence of the Neogene magmatic activity on this Mesozoic trapped ocean floor and thinned continental crustal slivers.

Origin of the volcanic islands of eastern Indonesia

We are concerned here only with the volcanic islands of the Banda (inner) arc and the Moluccas. The volcanic island chain east of Java that stretches through Flores and Wetar to Banda is the result of the northward subduction of the lithospheric plate that unites the Australia–New Guinea continent with the Indian Ocean. The oldest rocks of this arc have been dated as early Miocene (van Bemmelen 1949). The northward drift of Australia–New Guinea and the Indian Ocean from middle Eocene times (44 Ma) until early Miocene times (20 Ma), resulting from rapid spreading between Australia and Antarctica (Johnson and Veevers 1984), must have resulted in volcanic activity somewhere in south-east Asia. The apparent absence of volcanics of early Palaeogene age in the Sunda arc and the absence of volcanics of pre-Miocene age from the Banda Arc indicates that the present Banda subduction system was initiated in the early Miocene and suggests that during the early part of the 44 Ma to 20 Ma interval the boundary of the Indian Ocean with what is now the southern margin of Indonesia was a transform-type fault. This transform would have had an approximately NW–SE orientation if the post-Cretaceous anticlockwise rotation of Sundaland (Powell and Johnson, 1980) is correct. This suggests it was the change in orientation of the convergence between the Australia–New Guinea–Indian Ocean plate with Sundaland in the early Miocene (about 20 Ma) to a much more oblique angle that was related to the conversion of the transform into the Java trench subduction system.

The other volcanic islands we are concerned with here are those of the north Moluccas, viz. the Halmahera and Bacan group. Here there are very young volcanic islands of Quaternary age (Supriatna 1980; Apandi and Sudana 1980), but older volcanic rocks of Neogene and possibly also Palaeogene age are present in the main islands of Halmahera and Bacan. The shallow water that connects these islands to the north-west Vogelkop peninsula of Irian Jaya, and the presence of Benioff zones (McCaffrey 1983) east and west of this relatively shallow water region, suggest that Halmahera and Bacan have a basement whose affinity is with New Guinea. The most likely explanation for their present position is that they, together with Obi and the Banggai–Sula islands, and hence also eastern Sulawesi, have all moved west relative to New Guinea.

Origin of the non-volcanic islands of the collision zone

We are concerned here briefly with the origin of the non-volcanic islands of the collision zone (viz. Roti, Timor, Leti, Sermata, Babar, Tanimbar, Kai, Seram, Buru of the Outer Banda Arc as well as eastern Sulawesi) that are found between the Australian–New Guinea continent and the volcanic islands of the Inner Banda Arc (viz. Flores, Alor, Atauru, Wetar, Romang, Damar, Nila, Serua, Manuk, and Banda). There are two main theories of their origin: one that the islands have risen as a result of the sediments of the downgoing Australian plate being scraped off and piled up on the Asia side of the trench (Hamilton 1979), while the other view considers that the downgoing plate ruptured, overthrust the trench and carried the Australian margin northwards to collide with the volcanic Inner Banda Arc (Price and Audley-Charles 1983). There is probably agreement between both models that the main islands such as Timor and Seram began to emerge as islands in the late Pliocene (Kenyon 1974), and the emergence may have been a little earlier (early Pliocene) in Seram (Audley-Charles *et al.* 1979).

The Banggai and Sula islands expose Mesozoic rocks typical of continental shelf environments. Shallow marine deposits also characterize the Cainozoic succession. These observations and the present position of these islands as well as the imbricate zone at the boundary between Banggai and the eastern arm of Sulawesi all suggest the Banggai–Sula islands are a rifted Gondwanan continental fragment from New Guinea (Pigram and Panggabean 1984).

The eastern part of Sulawesi contains Mesozoic rocks very similar to those of the Outer Banda Arc islands of Timor, Seram and Buton. The Cretaceous strata appear to have been deposited in deep water on either a continental slope or proximal rise of the Australian–New Guinea continental margin. Sukamto and Simandjuntak (1983) have shown that the rocks of eastern Sulawesi were strongly deformed and overthrust, with thrusts moving towards the Banggai–Sula province, during the middle Miocene. This is generally interpreted to be the result of a collision in which the Gondwana fragment of eastern Sulawesi and Banggai–Sula collided with another Gondwana fragment of western Sulawesi. The western Sulawesi fragment is considered (Audley-Charles 1983) to have separated from Gondwana in the middle Jurassic whereas part at least of the eastern Sulawesi fragment was regarded as having separated in the early Jurassic (Pigram and Panggabean 1984) although I can find no evidence for it having separated from New Guinea until the Palaeogene, before its middle Miocene collision with western Sulawesi (Figs 2.8, 2.9). The collision suture is exposed in central Sulawesi. This collision may be compared with that of the Australian continental margin rocks of Timor colliding with the volcanic Inner Banda Arc in the mid-Pliocene. The geological evidence from Sulawesi in the form of erosional detritus from the deformed rocks suggests that the eastern part began to emerge as an island in the middle Miocene.

EVOLUTION OF LAND IN EASTERN INDONESIA

We know that there was much land exposed in Australia throughout the Cainozoic and that considerable tracts of land were also exposed then in Sumatra and in Malaya, Thailand, and Burma. In considering the distribution of land in eastern Indonesia during the Cainozoic we need to distinguish between the palaeogeographical picture before and after the collision between Australia and Asia.

Pre-collision land in eastern Indonesia

Eastern Indonesia was largely created by the collision. Before Australia–New Guinea collided with the Asian volcanic arc the only part of the region above sea level would have been the volcanic arc that stretched eastwards into the converging Indian Ocean. There are two aspects of this volcanic arc that deserve consideration. The volcanoes themselves appear to have been formed in the early Miocene (20 Ma), and some may have been above sea level at that time. When volcanoes cease to be active, as when the focus of eruption migrates, the volcanic edifices are rapidly eroded. It has been argued by Carter *et al.* (1976), Earle (1983), Brown and Earle (1983), and Price and Audley-Charles (1983) that the flat-lying thrust sheets in Timor (and by inference in other parts of the Outer Banda Arc) originated in the basement of the fore-arc to the Banda (inner) volcanic arc. Audley-Charles (1985) has pointed out how similar are the stratigraphy and structure of the post-Triassic allochthonous rocks on Timor to the exposed late Mesozoic to early Miocene section on Sumba. Thus before the collision between the Australian continental margin and the Banda volcanic arc it seems likely that a series of small low islands similar to Sumba were present immediately in front of the volcanic arc just as Sumba is at the present time (Figs 2.7, 2.8). Evidence from the overthrust rocks on Timor supports this suggestion. For example, the presence of shallow and very shallow marine Eocene limestones and very shallow marine limestones of early Miocene age in these thrust sheets, while not proving that there was any exposed land associated with these rocks does not discourage the suggestion, especially in view of the unlikely preservation of any terrestrial deposits after the profound erosion of these structurally high elements. Furthermore, the presence in Timor of an Eocene anthracothere (von Koenigswald 1967), which is a Tertiary relative of the hippopotamus, appears to support the suggestion of islands lying in front of the volcanic arc.

Post-collision land in eastern Indonesia

The collision of the Australian–New Guinea continental margin with the south-east Asian volcanic arc running through western Sulawesi and the Banda Arc resulted in the main features of the present configuration of eastern Indonesia (Figs 2.8, 2.9). Continuing convergence since the collision, together with much strike slip faulting associated with the general sinistral movement of Asia relative to Australia and the related acquisition of the sinuosity of the Banda Arc, have modified the initial pattern.

Essentially, islands have arisen where collision zones occur. This has resulted partly from crustal shortening and hence thickening of the low density crustal rocks, and the isostatic response to such thickening is uplift. In eastern Indonesia there has been both tectonic uplift of islands and sinking of land areas below the sea. For example, Audley-Charles and Hooijer (1974) pointed to the evidence of fossil pygmy stegodonts as indicating that the Wetar Strait between Timor and Flores had been uplifted above sea level by the collision process in the Pliocene but had sunk during the Pleistocene to below 3 km locally. Silver *et al.* (1985) have found evidence for the Banda ridges in the middle of the Banda Sea having subsided perhaps as much as 2 km during the Cainozoic. These ridges may have been above sea level during the early part of the Tertiary.

It is exceedingly difficult to determine how much of the present land area of Malesia was above sea level during the Plio–Pleistocene. What seems very probable is that the land area has substantially increased since the tectonic collision of Gondwanaland and the Asian arcs. Several processes may be identified as contributing to this increase in land area: the isostatic uplift of the collision zones, the eruption of volcanoes above sea level, the accumulation of sedimentary detritus producing a larger volume of rock than the parent being eroded, the accumulation of limestone reefs; all these processes have contributed to the increase in the land area of Malesia since the tectonic collision.

ROUTES FOR LAND PLANT DISPERSAL

The series of palaeogeographical maps (Figs 2.4–2.9) which have been based on the evidence and arguments developed in this chapter suggest that potential routes for land plant dispersal between Australia–New Guinea and mainland Asia came into existence in the Cretaceous as a consequence of the break-up of eastern Gondwanaland. This split off large slices of what were to become parts of south-east Asia, viz. Burma, Thailand, Malaya, and Sumatra. These regions remained for a large part above sea level during the late Mesozoic and Cainozoic. They became orientated NW–SE and so provided pathways connecting mainland Asia to Australia–New Guinea. The reorientation of this Asian region (Sundaland) by anticlockwise rotation during the Cainozoic resulted in the formation of volcanic arcs with which the northward-converging Australia–New Guinea collided. These collisions led to the formation of new land areas in the collision zones and provided new pathways for the dispersal of land plants between Asia and Australia.

ACKNOWLEDGEMENTS

I am much indebted to W. G. Chaloner, D. C. Christophel, J. Douglas, R. S. Hill and especially T. C. Whitmore for discussion and help with botanical literature, and to Janet Baker and Colin Stuart for the art work. P. Ballantyne and R. Hall transferred the qualitative geological reconstruction to the Cambridge Atlas World Map computer program and P. Ballantyne assisted in making the minor adjustments to fit the available magnetic anomaly data for the Indian Ocean.

3 FLOWERING PLANT ORIGIN AND DISPERSAL: THE CRADLE OF THE ANGIOSPERMS REVISITED

Armen Takhtajan

The earliest angiosperm fossils are still very scanty. They have been found from widely separated Lower Cretaceous deposits of mid-northern latitudes of Barremian and Aptian age (116–122 Ma) and are not particularly primitive. The cradle of the angiosperms and primary centre of their dispersal may have been far from the places of accumulation and fossilization, and cannot be revealed from our present knowledge of palaeobotany. Instead we must use the phylogenetic geography of the most archaic living forms. The greatest concentration of these is in south-west Asia and Melanesia, which includes north-east Gondwanaland. An origin in west Gondwanaland is considered less likely, for, although that region has suffered greater climatic perturbation and has a rich flora, it is much poorer in primitive Magnoliidae (described in detail) and also in Ranunculidae and Hamamelididae (given in outline). The east Gondwanaland concentration of archaic angiosperms is considered unlikely to be explained solely in the occurrence of optimum conditions for survival. The cradle and centre of diversification of the angiosperms was probably on some of the fragments described in Chapter 2 which rafted north from Australia–New Guinea in the Jurassic, perhaps as late as the Oxfordian (160 Ma). A mountainous archipelago is more likely than a single island to have provided the diverse habitats (including seasonally dry rain shadows) and the fragmented populations favourable for explosive evolution and adaptive radiation. Competition is less on islands than continents, and some mutants are likely to be more or less neutral which would be lethal on a continent. Thus, such an archipelago would also favour survival of the sort of neotenic macromutants ('hopeful monsters') which led to origin of the group. Long before their escape from the isolated area of origin the angiosperms must have undergone large-scale evolutionary transformations, to evolve their basic growth forms and ecological types. Expansion began around Neocomian time (144–125 Ma), to the tropics, subtropics, then later to cooler zones. The first emigrants were relatively advanced with better dispersal than the archaic forms; this fits the generally advanced character of the first fossils.

The fossil record of the oldest known flowering plants is still very scanty and is represented mainly by pollen grains and leaf impressions. Even so they are exceptionally important and demand careful examination.

Reports of angiosperm fossils from pre-Cretaceous rocks are erroneous or at least very dubious (see Scott, Bazghoorn, and Leopold 1960; Axelrod 1961; Takhtajan 1969; Wolfe, Doyle, and Page 1975; Hughes 1976; Hickey and Doyle 1977). Cornet (1980, 1981) reported the late Triassic 'monosulcate and polyaperturate angiospermid pollen', but unfortunately the publication is only in unillustrated abstract form and not yet described in detail. Besides, the absence of any authentic angiosperm megafossils from Triassic (and Jurassic) rocks considerably weakens the possibility of the angiosperm nature of this pollen.

More or less obvious angiosperm pollen first appears in the geological record in the Barremian strata of England (122 Ma) (Couper 1958; Kemp 1968; Hughes 1976, 1977; Hughes, Drewry, and Laing 1979; Laing 1976), in putatively correlative horizons in the basal Potomac Group of the Atlantic coastal plain of North America (Brenner 1963; Doyle 1969, 1973, 1977, 1978; Wolfe,

Doyle, and Page 1975; Doyle, Van Campo, and Lugardon 1975; Doyle and Hickey 1976; Doyle and Robbins 1977; Hickey and Doyle 1977; Walker and Walker 1984), and in equatorial Africa (Doyle, Biens, Doerenkamp, and Jardiné 1977).

The most striking and unexpected feature of the earliest known angiosperm pollen is its rather advanced character. Pollen grains of the Barremian angiosperms are noticeably less primitive than *Degeneria*-type pollen grains of extant 'lower' Magnoliidae. According to modern views on the evolution of the angiosperm pollen grain the most primitive type is monocolpate (monosulcate), boat-shaped, large- to medium-sized, more or less psilate (smooth), and atectate (non-interstitiate) (see especially Walker and Skvarla 1975; Walker 1976; Walker and Walker 1984). But there is no trace of this archaic type of pollen in the fossil record.* The early Cretaceous pollen grains are slightly boat-shaped to globose, medium-sized, reticulate and columellate. Therefore I agree with Walker and Walker (1984, p. 518) that *Clavatipollenites* and other currently known types of early Cretaceous angiosperm pollen grains represent relatively advanced pollen that 'is already too specialized to be able to reveal anything about the origin (or even the earliest evolution) of the flowering plants'.

Analysis of the megafossil record brings us to a similar conclusion. In Asia the oldest angiosperm leaf impressions are known from Transbaikalian Siberia and Primorie (Primorski Krai). A palmately trilobed leaf is reported (Kryshtofovich 1929; Krassilov 1967) from presumably Aptian (116 Ma) strata of south-eastern Primorie, and a very small entire pinnately veined leaf is reported from pre-Albian deposits of north-eastern Transbaikalia (viz. pre-113 Ma) (Vakhrameev 1973, 1981). In Europe the oldest leaves are known from the presumably early Albian of Portugal (*c.* 110 Ma) (Saporta 1894; Teixeira 1948). In North America the oldest leaves are reported from the probable Barremian–Aptian (120 Ma) strata of the Patuxent Formation of the Potomac Group (Fontaine 1889; Berry 1911; Doyle and Hickey 1976; Hickey and Doyle 1977; Doyle 1978). All these early Cretaceous leaves are already rather diverse and there is nothing particularly primitive in their morphology. Many of them have fairly modern organization of their venation, especially the Transbaikalian *Dicotylophyllum pusillum* Vachr. As regards 'poorly organized' venation, poor differentiation of vein orders and of the blade from the petiole (Doyle and Hickey 1976; Hickey and Doyle 1977; Doyle 1978), all these features can be seen in many advanced living taxa, both dicotyledons and monocotyledons. In comparison with all these Barremian and Aptian angiosperm leaves, leaves of the living genera *Circaeaster* (Circaeasteraceae) and *Kingdonia* (Circaeasteraceae) have much more primitive venation, although of secondary origin. In general the early Cretaceous leaves are even more advanced (and more diverse) than the contemporaneous pollen. They are even less able to reveal anything about the origin and evolution of angiosperms.

The vessel-bearing woods described by Stopes (1912, 1915) and attributed to the Aptian of England are also not particularly primitive. As regards the Aptian homoxylic wood known from Japan (Nishida 1962), there are some doubts as to the validity of its assignment to the dicotyledons (Wolfe, Doyle and Page 1975, p. 820).

Thus, the Aptian and even the Barremian flowering plants are already relatively advanced and there are no traces of the presence of any 'lower' Magnoliidae,* at least in pollen remains.

* There are in fact records of monocolpate, smooth, atectate pollen from the Jurassic and early Cretaceous. However, as Gingkos and Cycadophytes produced this type of pollen there is no inclination to attribute it to angiosperms, though clearly some could be of that source. (Editor)

* Within the flowering plants (angiosperms, Magnoliophyta) many *subclasses* can be recognized; amongst those believed to be primitive are Magnoliidae, Ranunculidae and Hamamelididae mentioned in this chapter. Every subclass contains several *orders*, each with several *families*. For example, subclass Magnoliidae contains order Magnoliales, which contains (amongst other families) Magnoliaceae, Winteraceae and Annonaceae. (Editor's note)

Hence we can conclude that the primary centre of the angiosperm dispersal may have been far from known areas of their accumulation and fossilization.

Evidently the flowering plants (Magnoliophyta) evolved long before the Barremian and most probably in pre-Cretaceous time. They could scarcely have attained such a grade of advancement and such a degree of diversification if they had arisen later than the latest Jurassic, at least the Jurassic–Cretaceous boundary. As Charles Darwin suggested in his letter to O. Heer (1875), angiosperms 'must have been largely developed in some isolated area, whence owing to geographical changes, they at last succeeded in escaping, and spread quickly over the world'. The 'isolated area' suggested by Darwin must have been far away from the Barremian and Aptian localities of fossil flowering plants. Otherwise it would be difficult to explain the absence of any representative of the 'lower' Magnoliidae among the Barremian fossils. It is very unlikely that only the relatively advanced forms were fossilized in or near the cradle of the angiosperms. The cradle was obviously far from the widely separated areas of the Barremian and Aptian angiosperm floras.

The clear conclusion therefore is that, unfortunately, the location of the cradle of flowering plants and primary centre of their dispersal cannot be revealed from our knowledge of palaeobotany. Therefore, for the present, any conclusions on the birthplace of the flowering plants are possible only on the basis of phylogenetic geography of their most archaic extant forms. 'If the fossil record of a group of organisms is poor or lacking, then its geographical history must be reconstructed (if at all) by inference from present conditions', says Simpson (1965, p. 77). 'In other words, only the result of a long sequence of events is known and the problem is to reconstruct the events from their results. That is always a hazardous and uncertain procedure . . . Yet some inferences can be made with fair probability provided that we can develop clear understanding of general principles effective in historical biogeography.'

The most archaic flowering plants (especially Magnoliales *sensu stricto*) are bradytelic (slowly evolving)* lines. Some of them, like *Degeneria*, are real living fossils, phylogenetic relicts, which survived from very remote times with only little change; others have undergone cladogenetic (diversifying)* evolution including wide speciation, but with little anagenetic (progressive)* change. The migration rates of the most archaic taxa are usually also relatively slower. Therefore there is a strong probability that the greater the geographical concentration of archaic taxa, the nearer is the centre of their origin. Of course archaic taxa may survive not only within or near the area of their origin but outside or even far removed from it. For example, the persistence of *Ginkgo biloba* in China does not mean that the genus originated in eastern Asia. Therefore there is always some danger in conclusions based solely on existing patterns of distribution, particularly on the distribution of systematically isolated taxa. But it would be much safer to base our conclusions on phylogenetic series of taxa. According to Komarov (1908, p. 380), 'General conclusions drawn from the geographical distribution of plants must be based primarily on the distribution of series, not on those of isolated species, for in these there is a much greater element of uncertainty and chance.' This idea of Komarov (we can name it the 'Komarov Principle') is completely applicable to the series of genera and even families. As a rule the most advanced members of the series are farther from the centre of origin. Therefore study of such series can give us an idea as to the directions and paths of their migration. If analyses of a sufficient number of phylogenetic series all produce similar results, we may consider it highly probable that our conclusions as to the centre of origin, and still more as to centre of dispersal, of a systematic group are near to the truth (Takhtajan 1969,

* editor's clarifications in parentheses.

p. 143). It is therefore important that the application of the 'Komarov Principle' to phylogenetic geography of most archaic flowering plants brings us to the conclusion that their primary centre of dispersal was in or near to that part of the Earth, which is now south-east Asia (including Malesia) and Melanesia (see Takhtajan 1957, 1961, 1969; Thorne 1963, 1976; Smith 1963, 1967, 1970).

South-east Asia and its adjacent areas form one of the most geologically complex regions, and some parts assigned to Laurasia were formerly parts of east Gondwana rifted from northern Australia–New Guinea. According to Audley-Charles, the rifting of south Tibet, Burma, Thailand, Malaya and Sumatra may have occurred 'as late as late Jurassic–Oxfordian time (160 Ma)' (Audley-Charles 1983, 1984, and this volume, Chapter 2). The cradle of the angiosperms was probably on one of the fragments rifted from northern Australia–New Guinea in the Jurassic. This hypothetical fragment must have been isolated from Laurasia and the rest of Gondwana until the early Cretaceous, when direct contacts or stepping-stone paths with nearby land masses were established.

The origin of the angiosperms in west Gondwana, postulated by Raven and Axelrod (1974) seems to me less probable. There is a considerable amount of evidence against west Gondwana as a possible primary area of angiosperm evolution.

In spite of the fact that the vast areas of west Gondwana (Africa and especially South America) have a very rich mesophytic flora, they are significantly poorer in archaic flowering plants than east and south-east Asia, Melanesia and Australasia. In fact, north-east Gondwana (present-day east Asia) has undergone much more geological and climatic perturbation than Africa or South America. It would, therefore, be difficult to explain why no representatives of the most archaic members of Magnoliidae have survived in the African flora, if they existed there previously. Of relatively archaic Magnoliidae only Canellaceae and Annonaceae are represented in Africa. In Madagascar there are noticeably more representatives of archaic families than in Africa: besides the Annonaceae and Canellaceae there are endemic genera *Takhtajania* (Winteraceae) and *Ascarinopsis* (Chloranthaceae). Although there are many more archaic magnoliophytes in America than in Africa and even Madagascar, there are still far fewer than in east and south-east Asia, Melanesia and Australasia; they are also represented in the former areas by more advanced taxa. Whereas there are ten endemic families of the Magnoliidae in east Gondwana countries (Degeneriaceae, Himantandraceae, Eupomatiaceae, Austrobaileyaceae, Amborellaceae, Trimeniaceae, Idiospermaceae, Rafflesiaceae *sensu stricto*, Nepenthaceae, Barclayaceae), there are only two endemic magnoliid families in America (Gomortegaceae and Lactoridaceae) and both are considerably advanced. Of the 12–15 genera of the family Magnoliaceae, only *Magnolia*, *Talauma* and *Liriodendron* (not counting a questionable northern South American endemic *Dugendiodendron*) are represented in America. The family Winteraceae, which contains eight genera, is represented in west Gondwana countries only by genera (*Drimys* in Central and South America, and the highly advanced endemic genus *Takhtajania* in Madagascar). The most primitive members of the Magnoliaceae and Winteraceae, as well as of the Annonaceae, Monimiaceae, Chloranthaceae, and some other magnoliid families are represented in east Gondwana countries. Phylogenetic–geographical analysis of the Ranunculidae, Hamamelididae, and other relatively archaic subclasses brings us to a similar conclusion. It is therefore exceedingly unlikely that west Gondwana was the primary centre of angiosperm evolution, and it is equally unlikely that the concentration of such a large number of archaic flowering plants in the east Gondwana countries can be explained solely by the presence in this region of optimum conditions for survival of archaic taxa.

The initial diversification of flowering plants and possibly their very origin could probably take place on a fragment (a 'shard'—Whitmore 1981) of east Gondwana (perhaps a fragment of the mysterious 'lost Pacifica continent', see Nur and Ben-Avraham 1977; Audley-Charles 1983) which was isolated over a long period of time. It was probably an archipelago or island arc rather than a single island.

An archipelago (especially a mountainous archipelago), with its wider spectrum of ecological opportunities, including for example the possibility of perhumid (rain-relief) and seasonally dry (rain-shadow) climates, and the barriers between islands, is a more favourable environment for 'explosive' evolution and diversification than a single island. This 'archipelago effect' (Carlquist 1965, p. 62) manifests itself both in rapid adaptive radiation and in intensive non-adaptive random genetic drift. Insular biotas (especialy those of oceanic islands) have a lower level of competitiveness (see Carlquist 1974) and therefore islands are more favourable for saltatory macroevolution and allow a greater free space (the '*patio ludens*' of Van Steenis 1969, 1977) than do continental environments. 'Some mutations that would be lethal or disadvantageous in continental environments have a more nearly neutral value in the less competitive environment of an oceanic island', says Carlquist (1974, p. 33). This less competitive environment favours not only the survival of prospective macromutants but also their further rapid evolution and diversification.

Many years ago I hypothesized that the angiosperms were probably the product of a neotenous* transformation of some unknown cycadophytic gymnosperm ancestor (summarized in Takhtajan 1976). Later (Takhtajan 1983) I suggested that the original flowering plants were probably a kind of Goldschmidtian 'hopeful monster' (prospective macromutants). They were neotenic 'monsters' juvenilized both in their vegetative and reproductive structures.

The neotenic precursors of flowering plants most probably arose in a region with a moderate seasonal drought (Takhtajan 1957, 1961, 1969; Axelrod 1970; Stebbins 1974; Doyle and Hickey 1976; Hickey and Doyle 1977; Doyle 1978). Doyle (1978, p. 386) suggests, that the angiosperms originated 'in unstable, "r"-selective environments', which agrees with the hypothesis of their origin in a variable seasonal climate. Although these inferences are highly conjectural, they enable us to explain the origin and ecological significance of angiospermy. The reduction, simplification and aggregation of gymnosperm micro- and megasporophylls into an angiosperm flower, the origin of the conduplicate carpel from an open gymnosperm megasporophyll, the origin of reduced and rapidly developing minute ovules, the origin of simplified male and female gametophytes without gametangia and of 'double fertilization', as well as features of their vegetative organization and rapidity of development and reproduction are best explained by neotenous transformation of gymnosperm ancestors under ecological stress in an unstable environment. The closure of the megasporophyll would have had an adaptive value in a seasonally dry climate both for protection from drought and potential insect predators and for rapid and easy growth of the pollen tube (Stebbins 1974). As a result of neotenic transformation of the gymnosperm megasporophyll, the conduplicate angiosperm carpel acquired the stigmatic marginal hairs, which was a very great improvement of the mechanism for cross-pollination. The stigmatic margins of primitive carpels, still preserved in some archaic living magnoliids (notably in *Degeneria* and *Tasmannia*), gave rise to a more organized typical stigma (at first decurrent and later terminal), which is one of the most

* With many other authors, both zoologists and botanists (including Stebbins 1974 and Corner 1976), I use the evolutionary term 'neoteny' in its broader meaning for any terminal abbreviation of ontogeny and premature completion of development of the whole organism (sporophyte or gametophyte) or any parts of it, that is for any hereditary morphological juvenilization (the shift of juvenile characters towards late ontogenetic phases). This 'Peter Pan' evolution includes both Kollmann's neoteny (neoteny *sensu stricto*) and Giard's and Gould's progenesis (see Gould 1977) as two different modes of hereditary juvenilization.

important parts of the angiosperm syndrome.

Long before the 'escape' of flowering plants from the 'isolated area' of their origin, postulated by Darwin, they must have undergone rapid and large-scale evolutionary transformations and became adapted to different altitudinal belts and various environmental gradients. They must already have evolved their basic growth forms and ecological types including deciduous woody dicotyledons, aquatic plants, xeromorphic plants adapted to dry climate, etc. Therefore, when in the Neocomian (144–125 Ma) the angiosperms began their expansion they could migrate along different geographical and ecological pathways and colonize different climatic zones. At first they colonized tropical and subtropical regions, and only later they moved towards the cooler parts of the Earth. The first emigrants were most probably relatively advanced taxa with greater dispersal abilities than the archaic forms, which are usually characterized by low dispersibility. This inference is in agreement with the geological record, which indicates the general advanced character of Barremian angiosperm fossils.

4 THE AUSTRALIAN–SOUTH-EAST ASIAN CONNECTION: EVIDENCE FROM THE PALAEOBOTANICAL RECORD

Elizabeth M. Truswell, A. Peter Kershaw, and Ian R. Sluiter

Traditional concepts concerning the origin of Australia's flora have involved the migration into the continent of floristic elements from both south and north. The development of plate tectonic theory stimulated this idea; it stressed that there had been a long period of biological isolation of Australia in the Cretaceous and early Tertiary, followed by the intrusion of Indomalesian taxa within the last 10 to 15 Ma, as a consequence of the collision of Australia with south-east Asia. Increasingly, detailed examination of the pollen records challenges this view that Australia was biologically remote from land masses to the north prior to the late Tertiary collision events.

In analysing the fossil record, it is necessary to distinguish between the origin of the angiosperms as a group and the origin of the lineages which may have given rise to extant taxa. The palynological record now available suggests that the earliest angiosperms appeared in Australia in the Albian, which is some 10 Ma after their appearance in other parts of the world. This gives little support to suggestions that continental fragments detached from the northern margin of Australia formed 'Noah's arks' transporting the earliest angiosperms northwards into Asia. Fossil data are so far inadequate to assess the route by which the earliest angiosperms entered Australia, but there is evidence to suggest that they developed in coastal habitats.

It seems likely that some of the angiosperm families now represented in the Australian–south-east Asian region, including some of the apparently primitive families, may have had global, or pan-tropical distributions in the late Cretaceous and early Tertiary. Assessment of the known ranges in time of those extant taxa for which there is a fossil record in the region suggests that there was some measure of floristic interchange between Australia and regions to the north in the late Cretaceous and early Tertiary. There is limited evidence that dispersal occurred in both directions, with some Australian taxa crossing into Sundaland before the mid-Miocene collision. Continental fragments which may have been emergent in the region between Australia and the Asian mainland could have provided a stepping-stone route by which such exchanges occurred.

It is not possible, from the available fossil record, to identify any massive influx of taxa into Australia from the north following the collision. Although there may be some biases in the record, comparisons of pollen spectra from pre- and post-collision sequences suggest that there has been no major alteration to the composition and structure of Australian forests as a result of invading elements from Asia during the past 15 Ma.

INTRODUCTION

The concept of an Australian flora which owed its origins in large measure to invasion of the continent by plant groups which had evolved and diversified beyond its boundaries has become a firmly established part of Australian botanical tradition. Its origins lie in the 1860 essay of J. D. Hooker, which formed the introduction to his *Flora of Tasmania* (Hooker 1860). Hooker identified three elements in the Australian flora, viz. an autochthonous Australian element consisting of the essentially xeromorphic taxa of woodland and open forest, an 'Indo-Malayan' element comprising the taxa now growing in tropical and subtropical rain forest, and an 'Antarctic' element, which encompassed the temperate rain forest and alpine taxa.

Hooker's concepts coloured Australian botanical thought for more than a century, in spite of the fact that mechanisms for migration, and pathways available for it remained vague, depending in large part on ephemeral, poorly understood 'land bridges'. These concepts were developed in an era before the notion of mobile continents had gained wide acceptance. The idea that immigrant floras or taxa had had a major influence on the evolution of the Australian flora was given fresh impetus with the development, in the late 1960s of plate tectonic theory. From this, it became evident that Australia had severed its connections with Antarctica, and thence with the rest of Gondwanaland, by the early Tertiary, and had moved north to collide eventually with island arc complexes in the Indonesian area in the mid-Miocene. This scenario embodies a long period of isolation for the Australian biota, in particular a remoteness from contact with landmasses to the north, during much of the Tertiary. This isolation was allegedly broken after the mid-Miocene collision, and subsequent elevation of land areas led to the establishment of archipelagic pathways which permitted the invasion, through northeastern Australian gateways, of taxa from Malesia. At the same time as the plate tectonic setting for Australia was being elaborated, fossil evidence, much of it palynological, was accumulating which shed new light on the past distributions of many elements within the Australian flora. Some of that fossil evidence appears to challenge the notion of a long isolation of the Australian flora, demonstrating, as it does, the relatively early, 'pre-collision', presence in Australia of plants which occur in extant tropical floras, including those of south-east Asia.

While it is apparent that there is a growing body of fossil evidence for the Tertiary which points to an autochthonous, or at least a very ancient, origin in Australia for much of that continent's extant rain forest, and which argues against large-scale, post-Miocene invasion, the fossil evidence from the early Cretaceous suggests that the very earliest angiosperms in Australia were indeed of immigrant origin. Currently available evidence suggests that they appeared here some time later than they did in the contemporary tropics. This evidence disallows much of the still persistent speculation that Australia, or continental fragments detached from its northern margins, were part of the 'cradle of the angiosperms', a speculation which has been built on the presence of a diversity of apparently primitive angiosperms in the living rain forest of north-eastern Australia. The two issues, the origin of the angiosperms, and the origin of angiosperm-dominated rain forests, which may contain primitive angiosperm taxa, are clearly separate, and relate to events which are probably separated in time. They do, however, appear to have become confused in some contemporary literature.

In this chapter we provide an up-to-date account of the fossil evidence. This supports the idea of an early presence in Australia of many of the elements of living tropical forest floras including taxa now shared with south-east Asia. In order to set the Tertiary scene in perspective, and to provide a counter to arguments which claim angiosperm origins in Gondwanaland fragments originally part of the Australian margin, we provide a review of the Cretaceous evidence for early angiosperm evolution in the region. To provide an historical perspective, we have prefaced these reviews of the fossil data by an abbreviated account of the history of the invasion theory in Australian botanical thought. Throughout, the term 'south-east Asia' is used broadly, to encompass Malesia west of Wallace's line, as well as the adjacent continental regions.

THE INVASION THEORY IN AUSTRALIAN BOTANICAL TRADITION

The perceived need to invoke large-scale migration of floras to colonize the Australian continent is perhaps more difficult to understand than the somewhat vague mechanisms on which it depended, but as Werren and Sluiter (1984) have recently suggested, it probably has its roots in the essentially Eurocentric focus of nineteenth-century Australian science (see Seddon 1981). The persistence of the invasion theory, as first articulated by Hooker, for at least a century in Australian botanical thinking is clear from the precepts expressed in Nancy Burbidge's classic (1960) work on the phytogeography of Australia. Burbidge elaborated on the elements originally defined by Hooker, but stressed that the complexities of vegetation history make assignment of living taxa to these groups extremely difficult; for instance, taxa classified by Hooker as autochthonous may, she believed, have been derived from an early immigrant flora. By relating the phytogeographical data to contemporary geological thought, Burbidge was able tentatively to visualize several periods of active migration, both to and from Australia. The most important phase she considered to have been pre-Tertiary, considering that contact with south-east Asia was possible up to the late Cretaceous or early Tertiary. In the early Tertiary it was more limited, with a Timor–Celebes Trough considered to have constituted a barrier, but with pathways possible via an exposed Sahul Shelf and New Guinea. The final contact phase was a Pleistocene one when pathways via New Guinea were facilitated by lowered sea-levels.

There have been dissenters to the view that a large part of the living Australian flora is the result of invasion from elsewhere. Barlow (1981) and Werren and Sluiter (1984) have discussed Herbert's (1932, 1967, and references therein) opposition to the theory. He argued for the existence in Australia of an ancient palaeotropical flora established as long ago as the early Mesozoic. The modern floristic elements were, in his belief, derived by climatic and edaphic sifting of this ancient flora.

The advent of plate tectonic theory in the late 1960s provided a firmer geophysical framework against which the evolution of the Australian flora could be viewed. It clarified the concept of a former union of Australia with Antarctica as a part of Gondwanaland until severence in the late Cretaceous (although the degree of contact between the two continents from then until the formation of open seaway in the Eocene remains uncertain—see Cande and Mutter 1982). This geographical scenario, of Australia at the beginning of the Tertiary just recently severed from Antarctica, but well separated from India, Africa, and New Zealand, and lying in humid high latitude climates, provided the background against which a clearer concept of a flora of essentially Gondwanic aspect could be visualized. This flora, as described by Barlow (1981), included as autochthonous elements *Araucaria*, *Casuarina*, *Dacrydium*, *Nothofagus*, *Podocarpus*, Restionaceae, the Myrtaceae, and a diversity of Proteaceae. Fossil evidence, much of it palynological, which became available largely during the 1960s and 1970s, confirmed an ancient presence in Australia for these taxa, and confirmed too that the apparently tropical *Nypa* (Palmae) and *Anacolosa* (Olacaceae) and the tribe Cupanieae of Sapindaceae were ancient components of these high-latitude floras (see reviews by Martin 1978; Kemp 1978).

Differentiation of this apparently Gondwanan stock was the major force in shaping much vegetation, including that of the arid zone (Truswell and Harris 1982). It is clear that the application of plate tectonic theory led to an emphasis on the differentiation of much of this flora having occurred in isolation during most of the Tertiary. In Barlow's (1981) essay, and in papers by Nelson (1981), Walker and Singh (1981), and Smith-White (1982) there is clearly a view that the Australian flora was remote from contact with other floras from the time of separation from Antarctica until contact with the Sunda Arc in the mid-Miocene. Raven and Axelrod (1974) and Raven (1979) also laid great stress on the absence of contact with floras to the north in this interval, claiming that a gap of some 3000 km of chiefly open water separated Australia from continental Asia during much of the interval from the early Cretaceous until the Miocene. The 'collision' events of the mid-Miocene terminated the long era of isolation, and brought the Australian flora into contact, essentially for the first time, with the rich south-east Asian flora. The migration paths thus opened, via a now-emergent New Guinea, allowed the development in Australia of what Barlow termed the 'post-Miocene intrusive element', which is believed to have contributed significantly to tropical ecosystems. Nelson (1981) separated this intrusive element into a tropical sub-element of taxa mainly restricted to northern Australia, with relatives in south-east Asia, and cosmopolitan and 'neoaustral' elements.

In spite of this 'post-Miocene intrusive element' having been labelled, and defined in time, and, less clearly, in space, it remains extremely difficult to identify in either the literature, or the fossil record. No list of intrusive taxa was provided by Barlow. The interpretation of this element by Burbidge (1960) was based on degrees of endemism; many non-endemic genera in the Australian tropical flora have only a few species in Australia, and are hence suggested to be of Indomalesian derivation. The fact that it proved impossible to identify clearly an intrusive element in post-Miocene (probably Late Pliocene) pollen spectra from north Queensland, one of the 'portals' for entry of the invaders, led Kershaw and Sluiter (1982) to question its reality. In the sequence which they examined at Butcher's Creek, on the eastern edge of the Atherton Tableland, over 80 per cent of the pollen taxa are known to occur in the Oligo–Miocene sequences of the brown coals of the Latrobe Valley in Victoria, an area where geographical location and age make it extremely unlikely that any intrusive elements from the north could be present.

The difficulties surrounding the identification of the intrusive element have been one factor which has stimulated recent papers questioning the importance of mass invasion in the geologically recent past after a long period of isolation. Other factors underlying the challenge to what has become established theory are ecological, and concern the distribution and floristics of rain forest communities now growing in northern and eastern Australia. Webb *et al.* (1984) argue that disjunctions of rain forest community types which are identifiable at generic and specific levels suggest that there has been fragmentation of formerly extensive, heterogeneous rain forest vegetation. They find implausible arguments which depend on dispersal events which they claim are inadequate to explain the network of closely related and integrated rain forest ecosystems now present in Australia, and argue instead for a greater measure of autochthony in explaining their origins. Webb *et al.* (in press) list a number of ecological factors which militate against extensive long-distance dispersal of rain forest elements; these include the low vagility of many species, the limited possibilities for the establishment of such seed as may have been transported, and the obligate association of many taxa within rain forests, which would require their migration in unison in order to establish the integrated, widely scattered communities now discernible.

The apparent degree of autochthony of Australian rain forest vegetation which these arguments indicate led Webb and his co-workers to suggest that the terms 'Indo-Malesian' and 'Antarctic' be abandoned in describing the origins of tropical and temperate rain forests in Australia. The presence of taxa now shared between Australia and Malesia is best explained, they believe, in terms of ancient floral links extending back to the Cretaceous. The similarities between the vegetation of the two areas is, they note, more pronounced between comparable habitats; they are more clearly recognizable in dry monsoonal forests than in wet rain forest. The explanation which they prefer to explain floristic affinities between the two areas, namely the joint inheritance of archaic stocks, is essentially that of vicariance biogeography.

THE EARLY ANGIOSPERMS IN AUSTRALIA

Takhtajan (1969) used patterns of distribution of a number of allegedly primitive angiosperm families to suggest that the tropical western Pacific region was the centre of origin for the angiosperms. His claim that the 'cradle of the angiosperms' was to be found somewhere 'between Assam and Fiji', perhaps in south-east Asia *sensu stricto* was influenced by the observation that the most primitive extant angiosperm families, and the most primitive genera within those families, are restricted to, or concentrated in, the islands and bordering continents of the western Pacific. Such geographical concentrations were cited for the Amborellaceae, Austrobaileyaceae, Calycanthaceae, Degeneriaceae, Eupomatiaceae, Himantandraceae, Lardizabalaceae, Magnoliaceae, Tetracentraceae, Trochodendraceae, Winteraceae and others. Takhtajan's claim was qualified only marginally: 'it was this part of the world which had been, if not the birthplace, then at least the original centre of the widespread Cretaceous expansion of the angiosperms; and this could hardly have been very far from their birthplace' (Takhtajan 1969, p. 142).

As an aside, it is pertinent to note that the alleged primitiveness of some of the taxa used in this argument need not link them back to a source in time close to the origin of the first angiosperms. The comment made by Walker (1976*a*, p. 286) is cautionary: 'Due to mosaic evolution, primitive taxa are rarely ancestral. A primitive taxon is simply one that retains a large number of primitive characters relative to some other taxon. Such retention of primitive characters is frequently used to infer a comparatively early evolutionary origin, but this may not always be the case.'

The lack of a fossil record from south-east Asia for the early Cretaceous, which is close in time to the origin and initial radiation of the angiosperms, has always constituted a weakness in claims for a south-east Asian origin. Other workers (Hughes 1976; Raven and Axelrod 1974) rejected Takhtajan's claim, favouring instead the idea that south-east Asia and the western Pacific may be a refugium for survivors of what may have once been a widespread tropical flora. Hughes's suggestion was that the flora may have been one that spread along the north shore of Tethys in the early Tertiary; this suggestion was based on the presence of the high proportion of taxa with south-east Asian affinities in the Eocene floras of Europe, especially in the London Clay flora.

Claims for an origin of the angiosperms in south-east Asia have been revitalized with suggestions by Takhtajan in the present volume (Chapter 3), that the centre of angiosperm evolution may have been on continental fragments rafted from the northern Australian margin into south-east Asia. According to available geological data, such rafting of continental blocks did not occur after the late Jurassic (Audley-Charles, this volume, Chapter 2; Pigram and Panggabean 1984).

Had such continental masses been central to earliest angiosperm evolution, then it might be anticipated that sedimentary sequences in northern Australia would contain fossil evidence reflecting this. Such evidence appears to be missing, even after the extensive palynological investigations of the past twenty years. These studies on early Cretaceous sequences in north-eastern Australia, based on analysis of scores of sections in, for example, the Eromanga and Carpentaria basins, indicate that the first angiosperms appeared here considerably later than they did in contemporary tropical and north temperate areas (Burger 1981; Dettmann 1981).

Fossil evidence for the early Cretaceous radiation of angiosperms (effectively summarized by Doyle 1977, 1978, 1984) shows clearly that they are unlikely to have had a lengthy pre-Cretaceous history. The use of pollen data invalidates earlier claims (e.g. Axelrod 1959) that the absence of angiosperms before the Cretaceous was due to their development in upland habitats: pollen from such habitats should be represented in lowland basins. The palynological record shows two successive phases in the appearance of pollen morphotypes. First to appear widely, in Barremian (125 Ma) or slightly older sequences, are grains with a monosulcate aperture and a columellate outer wall (but note the earlier, less complex morphotypes of Brenner 1984); this *Clavatipollenites* grain type bears some similarity to grains of extant Magnoliaceae, but Doyle

(1977) was careful to indicate that its presence in the early Cretaceous did not necessarily imply the presence of the modern family. Rather, he regarded the pollen morphotype as simply a grade of evolution through which pollen development may have passed. The monosulcate morphotype is widely succeeded by tricolpate pollen types. Evidence from sediments in Brazil and west Africa (in the northern Gondwana province of Brenner 1976) shows that this form diversified there in the Aptian (120 Ma); their times of appearance seem, on present evidence, to become successively younger in higher latitudes, confirming in a general way the suggestion made by Axelrod (1959) that the angiosperms originated within a broad tropical zone and from thence spread poleward.

In Australia, the oldest recognizable angiospermous pollen type is the *Clavatipollenites* morphotype. Its earliest documented occurrence is in the Early Albian of the Eromanga Basin in Queensland (Burger 1981), up to 10 Ma later than the Barremian or pre-Barremian (pre-125 Ma) occurrence in Africa and Europe. The monosulcate pollen in Queensland basins is succeeded, as elsewhere, by tricolpate forms (Fig. 4.1, from Truswell, in press) for which the earliest record is in the Middle Albian (*c.* 105 Ma) approximately coincident with their time of appearance in north temperate latitudes. By contrast, in the Cretaceous tropics of the northern Gondwana province, it appears at the base of the Aptian (*c.*119 Ma). Thus, the earliest records of angiosperms available from Australia are most appropriately interpreted as reflecting immigration of early angiosperm taxa, perhaps in success-

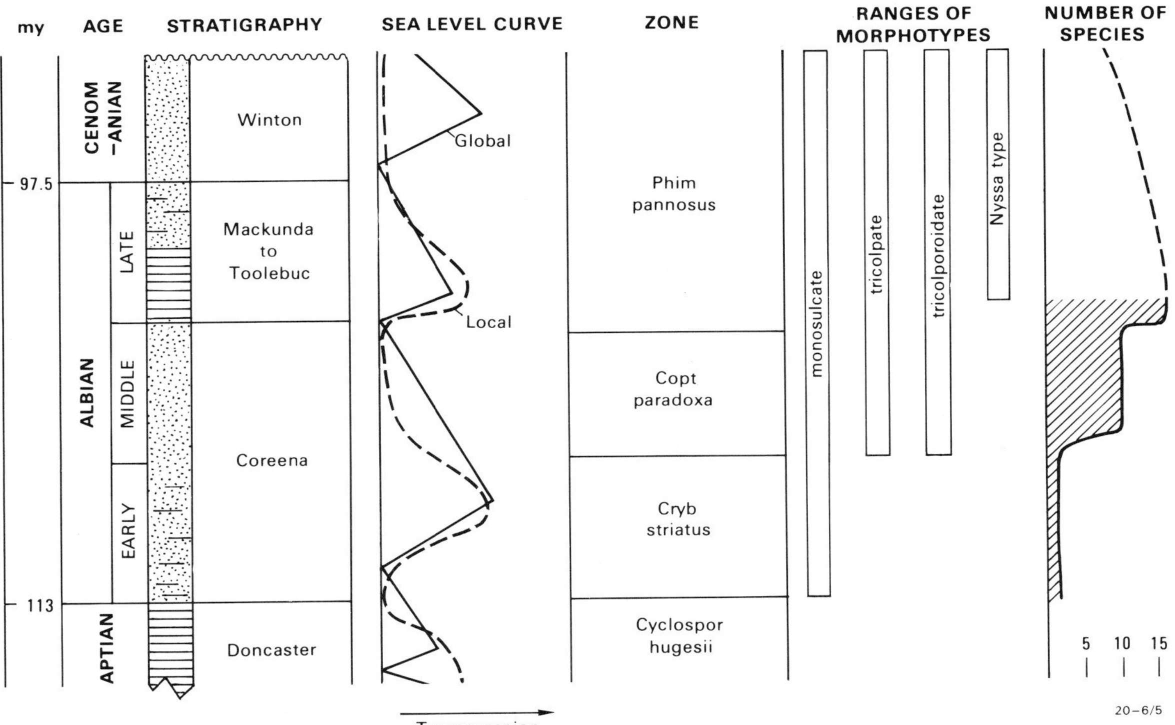

Fig. 4.1. Sequence of appearance of pollen morphotypes in Cretaceous sediments of the Eromanga Basin, Queensland. From Truswell (in press). Zones shown are adapted from the palynological zones of Dettmann and Playford (1969).

ive waves. Routes may have involved a southern shoreline of Tethys (broadly, a northern Gondwanaland margin) or, as favoured by Burger (1981, reproduced here as Fig. 4.2) a route along northern Tethys and thence southward through southeast Asia. There are insufficient fossil data to substantiate either option. Burger (1981), noting the apparent Aptian appearance of tricolpate pollen types in Japan (Takahashi 1974), suggested that a centre of development of the producers of these might have been present in southeast Asia, from whence they entered Australia (Fig. 4.2b). At present, the earliest monosulcate pollen type has been shown to make its first appearance in sedimentary basins in Queensland. It is, however, premature to interpret this as representing migration pathways from the north; much more detailed analysis of basins in

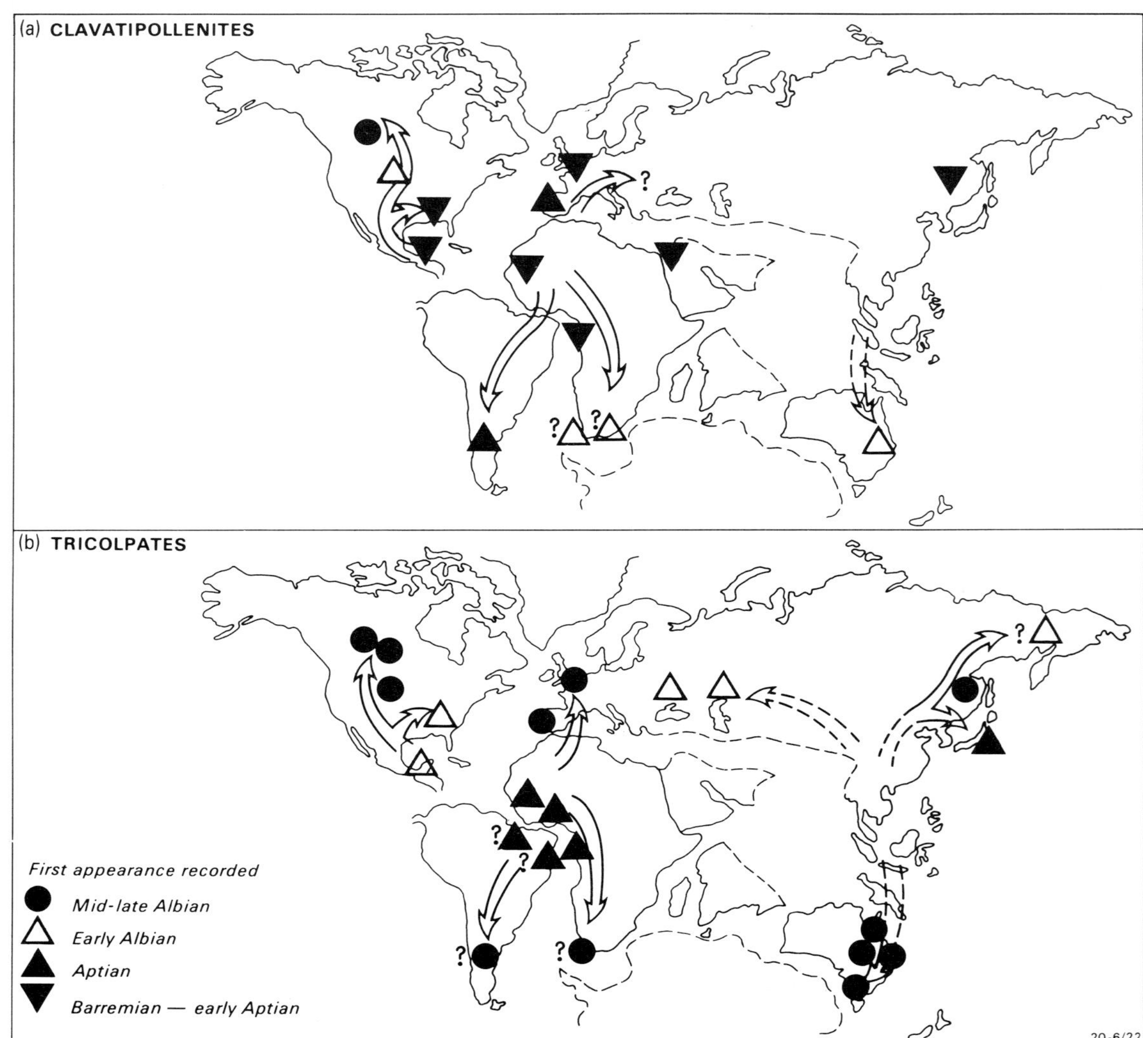

Fig. 4.2. First appearances of two angiosperm pollen morphotypes in the early Cretaceous in different parts of the world. Arrows indicate possible migration routes (updated from Burger 1981).

the south-east is required before this can be established.

The possibility that the initial entry of the angiosperms into Australia was along coastal routes is supported by the relationship between angiosperm fossils and sediment type in Australia and elsewhere. From study of north American sequences, Doyle and Hickey (1976) deduced that the earliest angiosperms may have been streamside plants—shrubby, opportunistic colonizers of disturbed habitats. A similar conclusion was reached by Retallack and Dilcher (1981) who suggested that the expanded areas of coastal habitats made available by marine transgressions in the mid-Cretaceous may have provided accessible migration routes for early angiosperms. In Australia, statistical studies (Dettmann 1973; Burger 1980) have shown the highest densities of angiosperm pollen in the Albian to occur in near-shore sediments, suggesting a close link between the parent plants and shoreline habitat. From Doyle's (1978) overview of early angiosperm evolution, it may be that radiation into the adaptive zone represented by the forest canopy layer did not occur until some time after the initial radiation. A peak in the diversity of angiosperm fossils in the late Cretaceous (see Niklas *et al.* 1980) has been interpreted as representing accelerated evolutionary activity coincident with movement into that zone.

Thus it would appear that the first appearance of the flowering plants in Australia occurred well after their establishment and initial radiation elsewhere. The fossil record suggests that it is unlikely that northern Australia, or segments rafted from this region, were part of the 'cradle of the angiosperms'. Whether or not other emergent land areas 'between Assam and Fiji' could be assigned that role is doubtful, though not impossible, in the face of scarce fossil evidence. The claims for a centre of origin for the group in the African–South American tropical region of the northern Gondwanaland province (Doyle 1984) appear to have strengthened with increased fossil evidence from that area, but further evidence from other areas which were geographically tropical in the early Cretaceous is needed.

LATE CRETACEOUS–EARLY TERTIARY DISTRIBUTION PATTERNS

The problem of whether or not the general region from south-east Asia to northern Australia could have subsequently been a centre of radiation for angiosperm groups, perhaps including the lineages which eventually gave rise to those families now considered primitive, is an entirely different question from that of the origin of the group as a whole and deserves consideration. It is unfortunate, however, that the published fossil record for the late Cretaceous, which was perhaps a critical interval for the establishment of the lineages of many modern taxa, is almost non-existent for the region. The records of Muller (1968) for Sarawak and Brunei are the only published basic data. The extant taxa which he could identify were *Nypa*, Myrtaceae and Sapotaceae as well as dicolpate palmoid pollen.

In Fig. 4.3 we have plotted the localities from which published pollen data are available for the late Cretaceous and Tertiary. It is evident that not much can be said about relationships of the late Cretaceous floras of northern Australia with those of land masses to the north. The Cenomanian floras of Bathurst and Mornington islands (Norvick and Burger 1976; Dettmann 1973) are too old to shed much light on the relationships of extant floras, and are not marked on the figure. To assess properly interfloristic relationships in the area which is now the northern margin of Australia, we need detailed palynological studies of near-shore sections such as that in the offshore Bonaparte Basin, where late Cretaceous to Tertiary sequences can be dated independently by associated marine microfossils.

In order to illustrate the distribution in time and space of a number of flowering plants which are now shared between Australia and south-east Asia we have plotted in Fig. 4.4 known time-ranges for those for which there is a fossil record in Australia and/or south-east Asia. The world

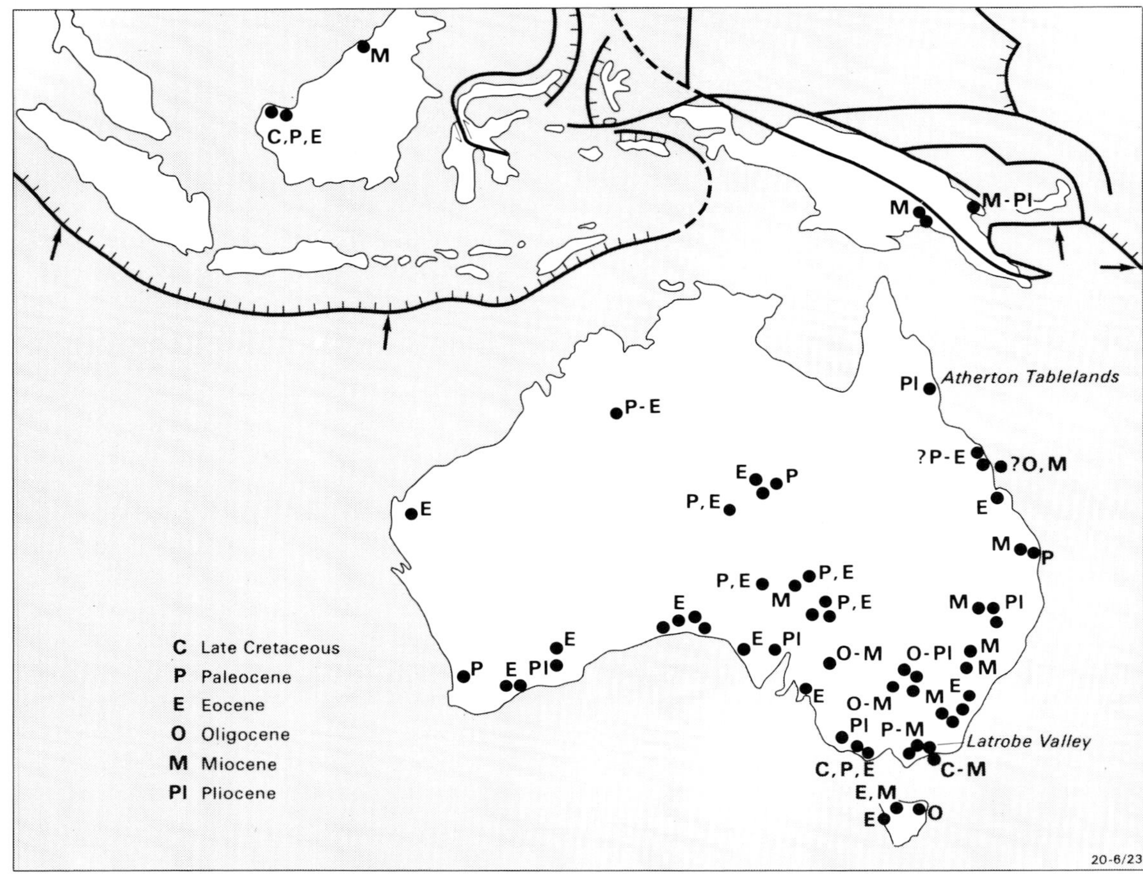

Fig. 4.3. Late Cretaceous and Tertiary palynological localities in the Australian–New Guinea and Malesian region.

time-range is also included where appropriate. It is evident that there are a number of categories of distribution. Firstly, there are forms which appear early in the Australian record which reflect distributions which were once global, or at least pan-tropical. Secondly, there are a number of forms which were, and have remained, exclusively southern. Thirdly, there are forms which have appeared earlier in Australia than they did in south-east Asia; and finally there are forms for which the reverse is true. It is apparent that there is a bias in the quantity of available data, with information for Australia being greatly in excess of that from south-east Asia.

There are some fossils, albeit very few, which suggest that the allegedly primitive angiosperm families, or the lineages which gave rise to these, may have been more widespread in late Cretaceous and early Tertiary. The Winteraceae is a clear example; its distinctive tetrads are present in the late Cretaceous of southern Australia and New Zealand, areas within the present geographical range of the family, but they have also been reported from the early Cretaceous of Israel (Walker *et al.* 1983), and the Tertiary of South Africa (Coetzee 1981; Coetzee and Muller 1985). Pollen very similar to that of Eupomatiaceae (whose sole living genus is confined to eastern Australia and New Guinea) has been reported too from the late Cretaceous of Cali-

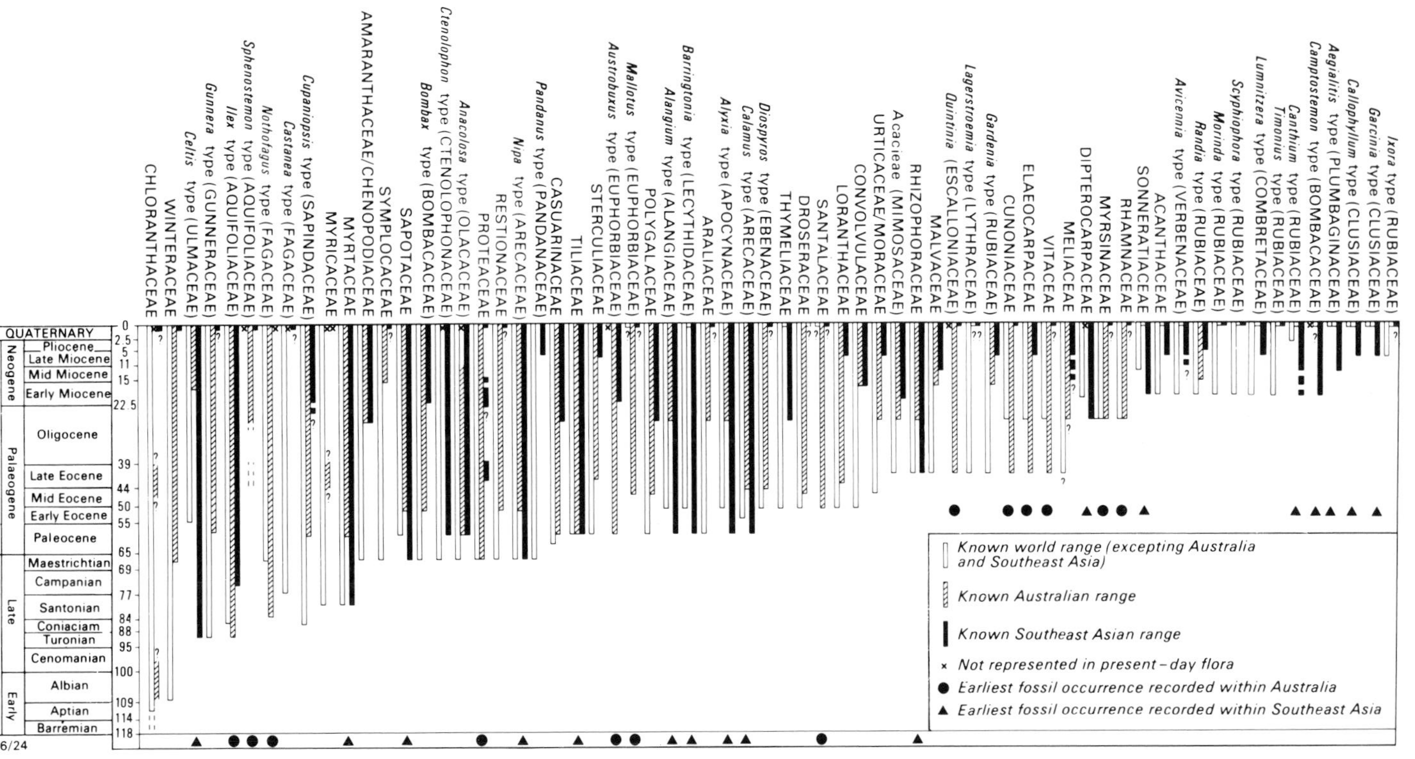

Fig. 4.4. Time-ranges of extant taxa represented in Australia and/or south-east Asia. Time-ranges on a world-wide basis are also shown for each taxon (updated from Muller 1981, including unpublished Asian data of I.R.S.).

fornia (Chmura 1973), but further records are needed to confirm this as indicating a formerly wider range; there are no fossil records from Australia. The record of Chloranthaceae is another which indicates a former wide distribution (Coetzee and Muller 1985), but, as stressed by Kemp and Harris (1977) and Muller (1981), it may be hazardous to assign early Cretaceous fossil pollen to the extant family.

Modern distributions which may be relictual from ancient pan-tropical distributions occur in groups other than the so-called primitive angiosperms. *Anacolosa* and related genera in the Olacaceae may be one such group. *Anacolosa* today has several species in Malesia and the western Pacific, and one in Africa–Madagascar; *Cathedra* occurs in Brazil, and *Ptychopetalum* in Africa and tropical America. This perhaps represents a relict pattern from a formerly more widespread angiosperm-dominated rain forest. The distinctive hexaporate pollen grains are reported in the late Cretaceous from Europe, Siberia, Japan, and California (Fig. 4.5a); the high latitude occurrences may be related to globally high temperatures in the late Cretaceous. In the Palaeocene–Eocene (Fig. 4.5b) the pollen type occurs in Europe, South America, Nigeria, Borneo, the Gulf Coast of North America, is widespread in Australia and New Zealand, and occurs at a number of localities on the Indian sub-continent (data from Muller 1981; Thanikaimoni *et al.* 1984; Martin 1982). Raven and Axelrod (1974) described the group as having had a 'west-Gondwanaland–Asian' history, and having migrated 'more-or-less directly' to Australia before the close of the Cretaceous. Such an apparently direct route does not appear to have involved southern Africa, as there are no records of the pollen type in the diverse pollen suites there (Scholtz 1985). While it is probably premature to discuss migration routes in view of the poverty of fossil data from the late Cretaceous, it nevertheless does appear that interchange of the parent plants which produced this distinctive form could have occurred between Australia and south-east Asia in the late Cretaceous. Similar patterns may be evident for the *Cupaniopsis* type of the Cupanieae (Sapindaceae) whose former wide distribution has also been noted by Coetzee and Muller (1985). Fig. 4.4 shows the early Tertiary appearance in Australia of the group which was widely dispersed in the late Cretaceous. The oldest records, according to Muller (1981), are pre-Maestrichtian ones from tropical west Africa and India; latest Cretaceous records show an expansion in both North and South America. In Australia, the earliest record is from the Palaeocene, in the Otway Basin in the south-east (Martin 1982).

The ancestral lineages of the Myrtaceae may also have had a pan-tropical distribution. Muller (1981) reported that the oldest pollen was found in the Santonian of Gabon, the Maastrichtian of Colombo, and, significantly, the Senonian of Borneo. In Australia, the oldest published account is from the Palaeocene (see Martin 1982, Fig. 8). Raven and Axelrod (1974) speculated that the original differentiation of the subfamilies Myrtoideae and Leptospermoideae took place from an ancestral stock that occupied west Gondwanaland and Australasia. Myrtoidae in tropical Asia, they surmised, have recently been joined there by Leptospermoideae from Australia. The ancient links between west Gondwanaland and Australia and the Pacific, had possibly been via Africa, and had, according to Raven (personal communication in Johnson and Briggs 1981) been followed by extinction there. One problem for such an ancient distribution path is that pollen assemblages from southern Africa show no record of Myrtaceae in the late Cretaceous–Palaeocene interval (Scholtz 1985), nor are there any records from among diverse pollen suites from India (Muller 1981), an area which might reasonably be expected to have formed part of such a path. Myrtoideae do appear, on present pollen evidence, to have a relatively ancient presence in Australia—pollen of the *Acmena* type is reported in some abundance from Late Oligocene coals in the Latrobe Valley in Victoria (Sluiter 1984). Links between west Gondwana-

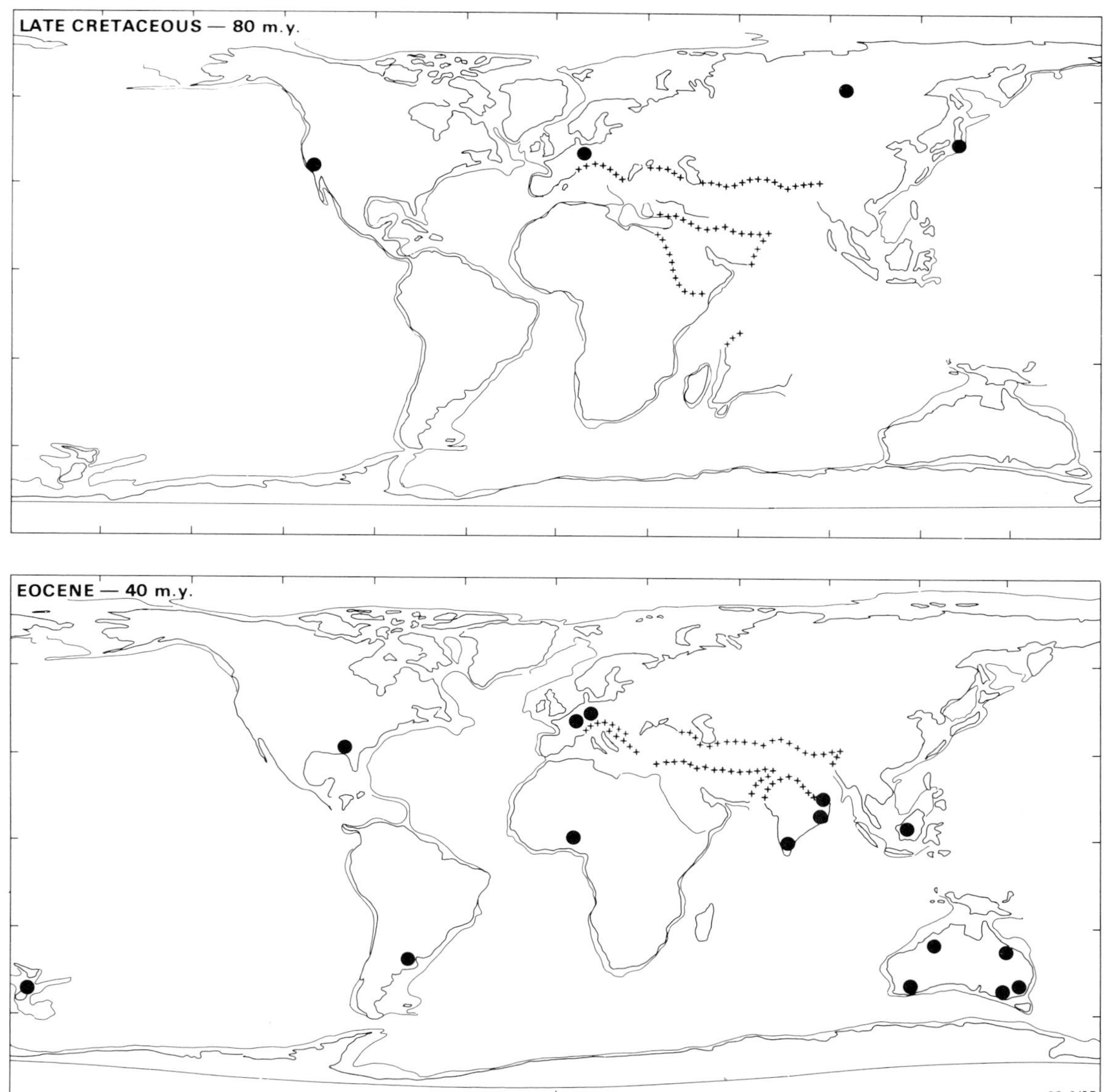

Fig. 4.5. Global distributions of *Anacolosa* pollen type. Late Cretaceous data from Muller (1981).

land and the Australasian area may have been via southern South America–Antarctica, since pollen of Myrtaceae certainly occurs in Antarctica (Truswell 1983) although the only *in situ* records are for the Eocene. However, given the documented early occurrence of the family in Borneo (Muller 1968, 1981), late Cretaceous–early Tertiary contacts between south-east Asia and Australia cannot be ruled out. Sapotaceae shows a similar pattern of a first appearance in south-east Asia, followed by a late record in Australia.

Among the palms, the mangrove *Nypa* presents the clearest picture of a former tropical distribution that was much wider than at present, as described by Dransfield in Chapter 6. The rattan palm *Calamus* is also interesting (Dransfield 1981, and this volume, Chapter 6). It occurs from Africa to Fiji with a strong concentration in Sundaland. There is Palaeocene and Eocene pollen from Borneo (Muller 1968) and recently, dicolpate, perforate, reticulate pollen closely similar to that of *Calamus* has been discovered in mid-Eocene sediments beneath dated basalts in the south-eastern highlands of Australia (E. M. Truswell, unpublished data). Firm identification is pending; if positive it hints that the *Calamus* lineage may have been a pre-collision entrant into Australia, probably from source taxa to the north, via the 'stepping-stone' islands.

A question that is frequently raised by biogeographers concerns the origin and differentiation of Fagaceae in the southern hemisphere. For example, Darlington (1965) proposed that ancestors of *Nothofagus* crossed into the southern hemisphere once, via the Indomalesian archipelago in the Cretaceous; Raven and Axelrod (1974) suggested that the group of Fagaceae ancestral to *Nothofagus* spread to Australasia from the montane tropics by the mid-Cretaceous, along topographical highs through Africa and India to Antarctica; and Whitmore (1981) hinted that differentiation of genera within the Fagaceae may have occurred on a landmass interposed between mainland Asia and Australia. The various themes were recently reviewed by Humphries (1981). There is abundant fossil pollen, but it reveals little of any dispersal route to the south. The early fossil records are confined to Australia, New Zealand and southern South America; in this Gondwanic region the group makes its first appearance in the Santonian epoch of the late Cretaceous. There are abundant records too from Antarctica, but as yet little chronological control on these. There is no evidence that the *Nothofagus* complex was ever in Africa or India; pollen suites from the late Cretaceous and early Tertiary in those regions contain no *Nothofagus*. Nor, from the limited data available, is there any evidence that it ever penetrated south-east Asia beyond New Guinea. Had *Nothofagus* differentiated from the parent Fagaceae within the south-east Asian region, as Whitmore (1981) suggested, then some record of it might be expected in the late Cretaceous to Eocene sequences of Borneo, described by Muller (1968), but there is none. The only records for New Guinea itself date from the Miocene (Khan 1974), by which time it was already well established. Questions of the derivation of *Nothofagus* within the Fagaceae, and of when the parent stock crossed the Equator, and indeed from which direction, remain unanswerable on present evidence. The available fossil data point to a great antiquity for *Nothofagus* within the Fagaceae; by contrast pollen of *Fagus* is first reported from the northern hemisphere in the Oligocene (Muller 1981). The fossil data therefore hint at a southern origin for the Fagoideae. A similar conclusion was reached by Humphries (1981) after a cladistic analysis of characters within extant *Nothofagus*. He was able to deduce only that the southern beeches are a wholly southern group whose origin is impossible to locate. From Fig. 4.5 it appears that *Nothofagus* is the only taxon which has remained so wholly southern in its distribution.

Much of the present discussion implies migration into Australia from regions to the north, but there is unpublished evidence that at least three taxa which were important elements in late Cretaceous and early Tertiary forests in Australia crossed into Sundaland at an earlier date than the classical Miocene collision. The Oligocene record of Casuarinaceae in south-east Asia may reflect one such event, and the sporadic appearance there of Proteaceae may reflect others. Jan Muller (in written communication to E. M. Truswell, 1983) indicated that the presence of the conifer *Dacrydium* (not shown on Fig. 4.4) suggested migration from Australia into Sundaland as early as 40 Ma (late Eocene); the presence of *Dacrydium* in China in the Oligocene has been established by Sun Xiang Jun *et al.* (1981).

LATE CAINOZOIC RECORDS

There are limited fossil data from northern Australia with which to determine the potential significance of a late Cainozoic invasion of taxa from south-east Asia. Only the stratigraphical sequences of Hekel (1972) from central and southern Queensland span the critical period from the Early to Late Miocene and from these there is no indication of an influx of new taxa.

Several taxa recorded from southeast Asia have been recorded later for Australia appearing there since the early Miocene (Fig. 4.4). These include *Celtis* type, *Pandanus* type, Acanthaceae, *Avicennia* type, *Canthium* type, *Camptostemon* type, *Aegialitis* type, *Calophyllum* type, *Garcinia* type, and *Lumnitzera* type. A number of these, however, are mangrove species which may have the capacity for long-distance dispersal and their distribution may not be constrained by the proximity of landmasses. The majority of these taxa are only first recorded between the Late Oligocene and Late Miocene so did not have to cross a significant ocean barrier to reach Australia. In fact, it is conceivable, given the lack of fossil evidence from northern Australia, that some actually migrated to south-east Asia from Australia.

There are a number of taxa on Fig. 4.4 which could more definitely have migrated to south-east Asia from Australia. These include *Gunnera* type, *Sphenostemon* type, Winteraceae, Symplocaceae, Sterculiaceae, *Mallotus* type, Araliaceae, *Diospyros* type, Droseraceae, Santalaceae, Loranthaceae, Urticaceae/Moraceae, Malvaceae, *Gardenia* type, Cunoniaceae, Elaeocarpaceae, Vitaceae, Myrsinaceae, Rhamnaceae, and *Randia* type. To this list can be added the southern conifers *Podocarpus* and *Phyllocladus*. In contrast to the taxa which are presumed immigrant to Australia most of these others were well established in Australia. Exceptions include *Gardenia* and *Randia*. *Randia*, which is now confined to north-east Australia, is the major stratigraphical zone indicator pollen taxon *Triporopollenites bellus* of the mid-Miocene in south-eastern Australia (Stover and Partridge 1973), and its rapid spread over such a large area gives the strong impression that it must have arrived in Australia not long before that time. Consequently, its recorded presence in Australia prior to south-east Asia is somewhat surprising. There are, however, a large number of Rubiaceae, including *Randia*, recorded first on islands in the western Pacific in the early Miocene before either Australia or south-east Asia and it is possible that this was a time and place of major diversification for them. The isolation of the western Pacific islands suggests again that the close proximity of landmasses need not be invoked to explain distributions.

An alternative method of assessing the extent of any invasive element, particularly with regard to its effects on the vegetation, is to compare quantitative pollen assemblages prior to and after the major contact phase. In a consideration of rain forest, the main period of potential contact between floras was likely to have been in the Miocene before increasing aridity reduced it to isolated fragments.

Pollen sequences from the Atherton Tableland, north-eastern Queensland, provide suitable data for the post-contact phase but it is necessary to turn to south-eastern Australia for the pre-contact period. Summarized sequences are shown on Fig. 4.6 The south-eastern Australian records are continuous brown coal sequences extending from the Late Oligocene to the Early Miocene (Sluiter 1984). The north-eastern Queensland data are made up of a short Plio-Pleistocene coal and oil shale sequence at Butcher's Creek (Kershaw and Sluiter 1982), a long Late Quaternary record from a volcanic crater swamp, Lynch's Crater (Kershaw 1985) and the last 2500 years, not represented at Lynch's Crater, from a second volcanic crater site, Lake Euramoo (Kershaw 1970).

Fig. 4.6 clearly shows that there are very close similarities between the pre- and post-contact sequences. All taxa except *Freycinetia*, *Trema* and *Lagarostrobus* occur in both sequences while many others, including Casuarinaceae,

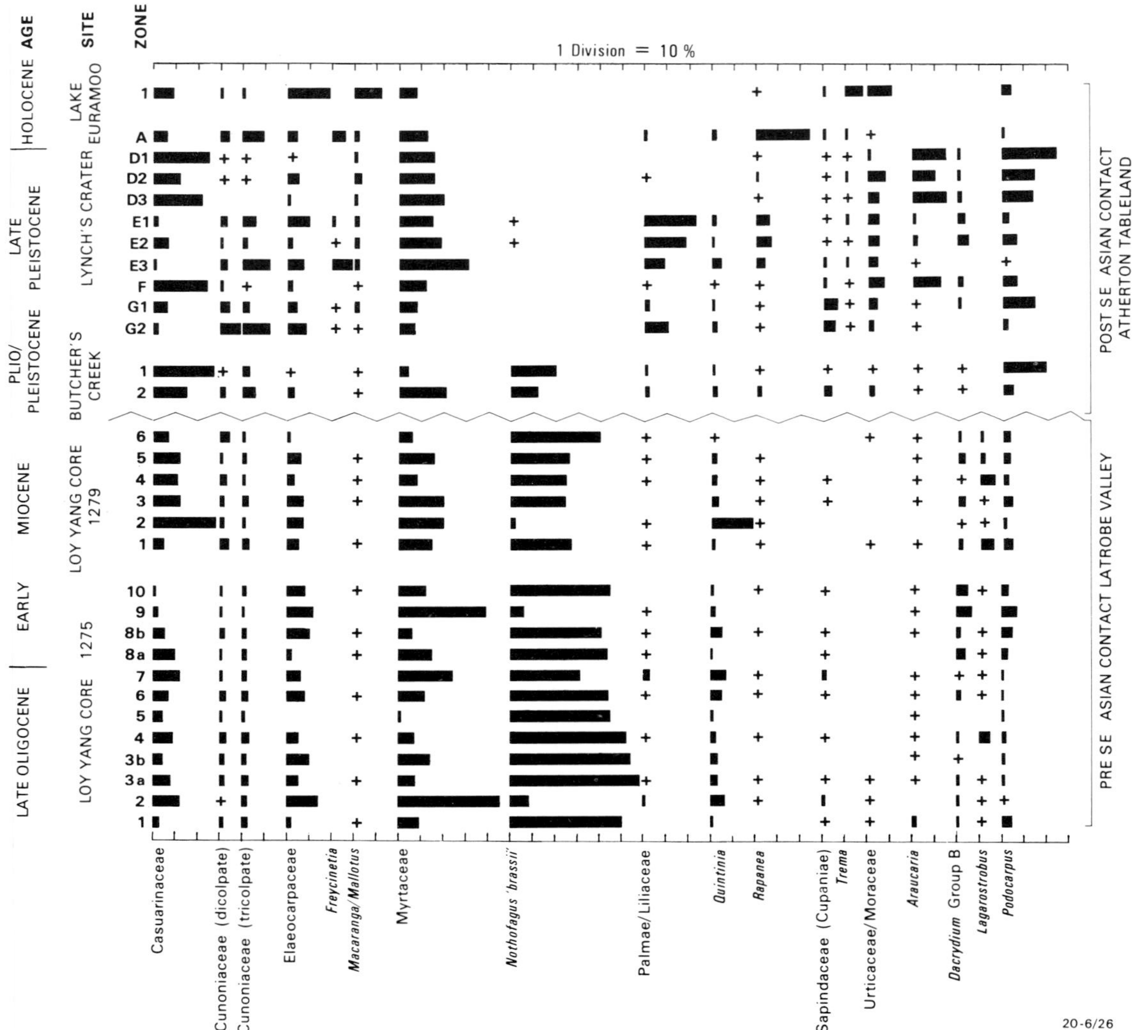

Fig. 4.6. Pollen spectra from selected pre- and post-Asian contact sequences in Australia. Average values are shown per taxon for each identified zone on the original pollen diagram, expressed as percentages of the forest woody-plant pollen sum of the zone. Only woody angiosperms and gymnosperms which achieve 5 per cent in any one zone are included.

Cunoniaceae, Elaeocarpaceae, Myrtaceae, *Quintinia*, Sapindaceae, *Dacrydium* Group B, and *Podocarpus* have reasonably comparable values between the two sequences. The lower values of *Nothofagus* and higher values of *Araucaria* in the north-east Queensland record reflect regional trends within the forest vegetation in response to climatic conditions rather than any influences resulting directly from continental movements. It is unlikely also that *Freycinetia* and *Trema*, represented only in north Queensland, migrated from south-east Asia in the late Cainozoic as neither has a fossil record there.

The general impression from Fig. 4.6 is that the influx of taxa from south-east Asia into Australia has been minimal and in fact there may have been a greater movement in the opposite direction. However, it must be borne in mind that these results are based on a rather limited, generalized and biased fossil record. One major limitation is the lack of specific taxonomic identifications and it is quite possible that interchange has occurred at undetected generic and, perhaps more importantly, at species levels. Another problem is that many taxa have not been recorded as fossils. This is particularly the case with tropical lowland rain forest plants which have limited, indiscriminate pollen dispersal, and is more likely to apply to taxa dispersing from the south-east Asian rain forests than from Australian forests. Despite these reservations, the fossil record does indicate that there has been no major alteration to the structure and composition of Australian forests as a result of any invasion of south-east Asian taxa within the last 15 to 20 Ma.

RECONCILIATION OF FOSSIL EVIDENCE WITH GEOLOGICAL DATA

The notion that Australia was separated from the southeast Asian mainland by some 3000 km of empty ocean to its north during the Cretaceous—a separation which allowed its flora to develop in splendid isolation—is rooted in the development of plate tectonic theory. The most popular palaeocontinental reconstructions, such as those of Smith *et al.* (1981, see also Powell *et al.* 1981) have been compiled from sea-floor spreading data and from palaeomagnetic information. The maps compiled are broadly drawn and do not show in any detail the former distribution of land and sea areas.

The complex geological history of the Indonesian region, is evident from a compilation such as that of Hamilton (1979). The possible interrelationships of this area with the northern margin of the Australian plate, significant in deciphering the former geography which is the backdrop to biotic exchange, have been outlined by Audley-Charles (1983, and this volume, Chapter 2) and by Pigram and Panggabean (1984). These authors visualize the rifting, eventual detachment and northwards movement of sections of the leading edge of the Australian plate. From the early Cretaceous, a complex of continental fragments existed between Australia and Asia which would have provided an adequate base for a kind of stepping-stone path for the exchange of plants and animals for much of the time since the early Cretaceous, especially in the late Cretaceous and early Tertiary. The maps in Chapter 2 suggest a decrease in potentially viable exchange routes in the Oligocene, followed by an increase after mid-Miocene time as a result of the collision then between the northward-moving Australian plate and island arc complexes.

Pigram and Panggabean (1984) make reference to a 'screen of micro-continents' being present off the northern margin of Australia since the late Jurassic. They identify a number of microcontinents which they consider to have been fragments detached from northern Australia–New Guinea, and suggest that these are now embedded in the tectonically complex terrains of Sulawesi and Irian Jaya (whereas Audley-Charles believes they lie further west). Their main concern is with identifying and dating possible rift sequences, and with establishing the age of the subsequent onset of sea-floor spreading. This latter event they date as early Jurassic in eastern New Guinea, and suggest that the age of spreading becomes successively younger to the west, so that the age of opening of the Indian Ocean off Australia's north-west, in the Wharton Basin, is mid- to late Jurassic, off its south-west margin early Cretaceous, and so on.

It is not our concern to attempt to assess the relative merits of differing models of geological evolution. We are concerned, however, with the implications for the palaeogeography of biotic exchange between Australia and Asia. In this concern we are hampered by scarcity of data. First, there are few published palaeomagnetic data which would allow the placement of purported continental fragments in their former latitudes. Data from Thailand, Malaya, and Borneo have been used to suggest that these regions have remained in essentially their present latitudes since the mid-Cretaceous (see references in Powell *et al.* 1981), but this has been challenged by Audley-Charles (1983). Second, there is a dearth of sedimentological and tectonic data relating to which fragments were emergent and available for colonization.

In Chapter 2 Audley-Charles considers the extent of exposed land in south-east Asia during the Jurassic and Cretaceous. He points to the widespread distribution of non-marine, often red-bed facies in the Jurassic and Cretaceous of South Tibet, Burma, Thailand–Malaya, and possibly Sumatra, suggesting that the facies may reflect exposed land that was available for colonization by plants. However, dating of these sequences is frequently uncertain, and unless the stratigraphical intervals concerned involve a significant mid- to late Cretaceous section, the exposed areas probably pre-date the major phase of angiosperm evolution. Some data relevant to former areas of exposure are contained in the stratigraphical sections illustrated by Pigram and Panggabean (1984, Fig. 10) for some of the microcontinents embedded in Irian Jaya. Coarse terrigenous sediments in these testify to emergent, eroding land areas in late Cretaceous; possible depositional hiatuses in some of the Cretaceous sequences may similarly reflect periods of emergence.

The general scenario of the position and degree of exposure of former land areas which is embodied in current geological models can, for the most part, be reconciled with the fossil data we have drawn together. We consider, however, that these data do not support the idea of the earliest angiosperms evolving on continental slivers rafted northwards from the Australian continental margin. This is ruled out by the time constraints of the fossil record; the earliest angiosperms appear in northern Australia, on current evidence, in the Albian, which is later than in other palaeotropical regions. The continental masses which lay between Australia and south-east Asia could have provided pathways, probably somewhat haphazard ones, by which means exchange of later developed angiosperms occurred. The fossil record suggests that there may have been a degree of interchange in the late Cretaceous and early Tertiary: this is clearly possible in the configurations suggested by Audley-Charles (this volume, Chapter 2) for 80 Ma and 40 Ma. The pathways provided by emergence of land areas following the mid-Miocene collision event would have been similarly haphazard, and of a stepping-stone nature. This is in clear accord with data from pollen spectra which indicate that no profound changes in Australian vegetation patterns followed this tectonic event.

SUMMARY AND CONCLUSIONS

The first angiosperms in northern Australia arrived there later than they did in other parts of the world, in particular western Gondwanaland, an area which has some claims as a centre of angiosperm origin. Colonization in Australia may have been along coastal pathways, but the routes by which the first angiosperms reached Australia remain unknown. The suggestion that continental fragments rifted from northern Australia in the late Jurassic acted as Noah's Arks (see McKenna 1973 for discussion of dispersal methods) transporting ancestral angiosperms northwards into Asia is not borne out by fossil evidence. The fossil record for Australia suggests that such fragments were unlikely to have carried angiosperms. At the close of the Jurassic Australia seems likely to have borne a flora dominated by gymnosperms with Araucariaceae and Podocarpaceae (Gould 1976; Dettmann 1981).

While, on present evidence, it does not seem likely that the fragments rifted from northern Australia carried an angiosperm flora at the start of their journey, they may well have become part of a stepping-stone route by which means later developed angiosperms were exchanged between Australia and south-east Asia. Emergent land areas perhaps formed an erratic 'sweepstakes' route across which a degree of floristic exchange may have occurred prior to the mid-Miocene collision. These migrations took place as individual occurrences within wider distribution patterns. Limited pollen data now available suggest that some angiosperms, including some of the allegedly primitive ones, may have had global, or at least pan-tropical distributions in the late Cretaceous and early Tertiary. Examples are Winteraceae, Chloranthaceae, the *Anacolosa* lineage, some Sapindaceae and Myrtaceae, and some palms. Although it is often assumed that migration into Australia from the north has predominated, there is some evidence also of northwards migration of late Cretaceous–Tertiary floras into Sundaland prior to the mid-Miocene collision; examples are Casuarinaceae, *Dacrydium*, and certain Proteaceae. The stepping-stone route, while providing a passage for exchange of some taxa, seems also to have acted as a filter preventing penetration of others, notably *Nothofagus*, into Asia.

It is not possible, from the fossil record currently available, to identify any massive influx of taxa into Australia from south-east Asia since the mid-Miocene. Conceivably some invasive elements remain undetected in the record, which has some biases. What is apparent from the pollen record is that there has been no major alteration to the composition and structure of Australian forests as the result of invasions during the past 15 Ma.

The accumulated fossil data challenge the concept of an isolation of Australia that was prolonged and intense during the Cretaceous and Tertiary. There is increasing evidence that there was some taxonomic interchange with regions to the north as long ago as the mid-Cretaceous. It should be stressed, however, that the data we have been able to present here represent only a beginning. To assess adequately the interrelationships of floras in south-east Asia there is clear need for much more published palaeobotanical information. A great deal exists in oil company files: the release of just some of these data would do much to clarify past and present floristic relationships of land areas within southeast Asia.

ACKNOWLEDGEMENTS

EMT publishes with permission of the Director, Bureau of Mineral Resources. We thank Chris Pigram for discussion of Indonesian geology, and Helene Martin and John Flenley for critically reading the manuscript.

5 LATE CAINOZOIC VEGETATIONAL AND ENVIRONMENTAL CHANGES IN THE MALAY ARCHIPELAGO

R. J. Morley and J. R. Flenley

A brief review is given of evidence for changes in sea levels, degree of seasonality in precipitation, and variations in temperature during the late Tertiary and Quaternary in the Malay archipelago. Also new evidence is presented for a seasonal climate within the Malay peninsula during the middle Pleistocene.

Tentative palaeoclimatic maps are constructed for Quaternary glacial maxima and for the early Miocene. These reconstructions reveal that during glacial maxima, and at times during the late Tertiary, seasonal climates were more extensive in the Malay archipelago, allowing migration of 'savanna' plants and animals across the area. During glacial maxima cooler temperatures lowered montane vegetation zones providing more stepping-stones for mountain taxa to migrate. Late Tertiary and Quaternary palaeoclimates thus help to explain many existing disjunct plant and animal distributions in the area.

INTRODUCTION

The present distributions of living organisms in the Indo-Malesian region are of many types. Apart from the extremes of ubiquity and endemism and the classical distributions limited by Wallace's line (Whitmore 1981), there are two types of fragmented distribution which are of special interest. One particularly significant type is the disjunct distribution of plants of lowland seasonal climates between continental Asia and south Malesia. A classic example is *Rynchosia minima* (Leguminosae) (Fig. 5.3(b) in Whitmore 1981). Another example is the genus *Aegialites* (Plumbaginaceae), which occurs in both Indo-China and Australasia (Fig. 5.1). Both these regions experience a long dry season, in contrast to the ever-wet climate of Borneo, Malaya and Sumatra. Similar distributions are exhibited by many extinct mammals, a few living mammals (Medway 1972), and by many plants. A second significant type of fragmented distribution is exhibited by many mountain peak taxa. Such disjunctions may be massive, as in *Drapetes ericoides* (Thymelaeaceae), *Euphrasia* (Scrophulariaceae) or *Primula prolifera* (van Steenis 1934–36), or at a more local level—for instance, the genus *Astelia* (Liliaceae) occurs on several high peaks in New Guinea, but is absent in between (Fig. 5.2). Clearly this distribution is particularly related to the availability of suitable open habitats above the forest limit. Further disjunct distributions of tree species are described from Java by van Steenis (1972), e.g. *Albizia lophantha* (Leguminosae) and *Myrica javanica* (Myricaceae).

Although these distribution types are, in some ways at least, in accordance with the modern environment, the question of how a single species came to have one or more disjunctions in its range invites historical explanation. Apart from the major geological changes already discussed by Audley-Charles (this volume, Chapter 2), the kinds of historical event which would usefully contribute to an explanation are environmental changes of three main types, namely sea-level changes, changes in the degree of seasonality in precipitation, and temperature changes.

A reduction of sea level would provide land linkages from mainland south-east Asia and Australia out to the islands. An increase in seasonality might make these linkages available to organisms requiring a seasonal climate, thus

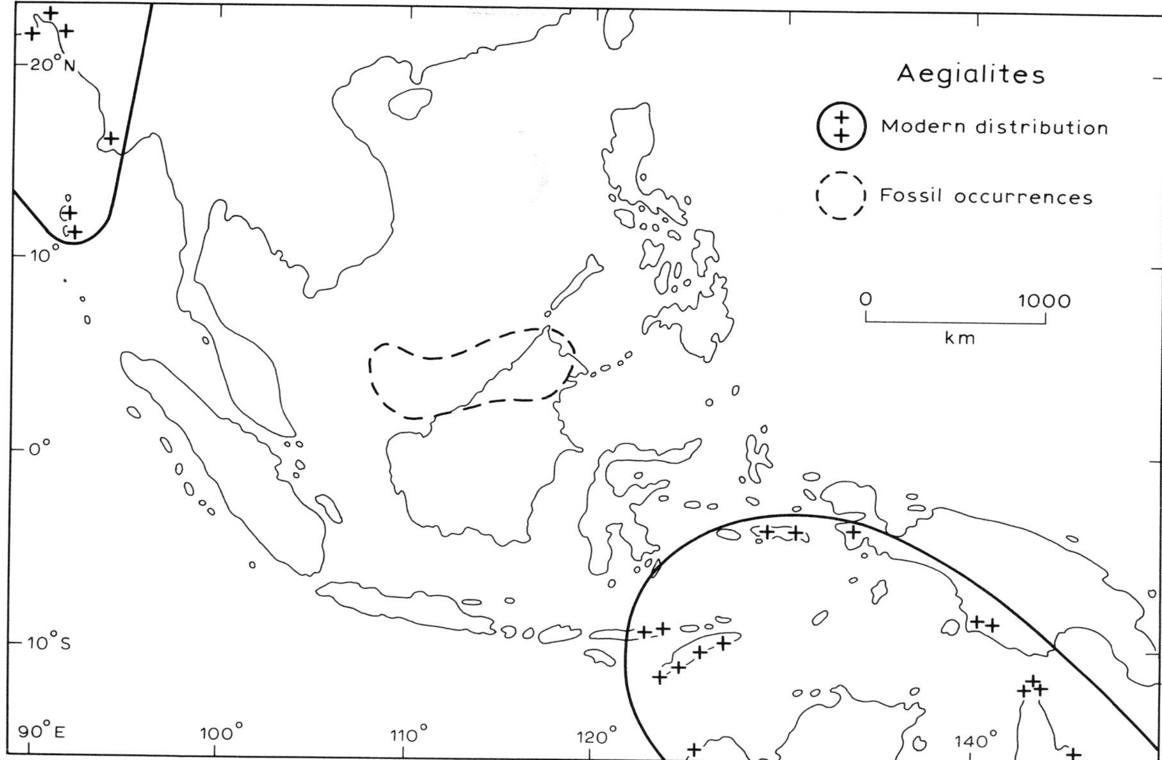

Fig. 5.1. Distribution of the genus *Aegialites*. (After van Steenis 1949.)

assisting migration between, for instance, Thailand and Sulawesi. A reduction in temperature would lower the montane zone boundaries, thereby making lower hills available to stenotherm (lower temperature) species, and thereby aiding stepping-stone migration.

In recent years the complex world-wide environmental changes of the late Cainozoic, most conspicuously manifested by a repeated alternation of Glacial and interglacial periods, have become more clearly understood, and there is now considerable evidence for a single unifying model which is illustrated, in a generalized way, in Fig. 5.3. It now seems probable that the major climatic cycle, with a period of *c.* 100 000 years, is initiated by astronomical forces, most likely originating from small changes in the ellipticity of the Earth's orbit, which in fact has a period of about 100 000 years. The changes in insolation resulting from this ellipticity are, by themselves, however, too small to cause glaciation; there must therefore be some amplification by positive feedback, to enlarge these changes. Two such amplification mechanisms are now appreciated, one operating world-wide, the other only in polar and near-polar regions. The world-wide mechanism is the change in CO_2 concentration of the atmosphere. The fact that such a change actually occurred has now been clearly demonstrated (Shackleton *et al.* 1983). Since CO_2 in the atmosphere causes the well-known 'greenhouse effect', lowered CO_2 at glacial maxima may have had a profound influence on world temperatures. The second reinforcing mechanism is the influence of albedo. The increase in albedo of ice-covered surfaces causes decreased absorption of insolation in glaciated regions, explaining why temperate climates fluctuated more than tropical ones in the Quaternary.

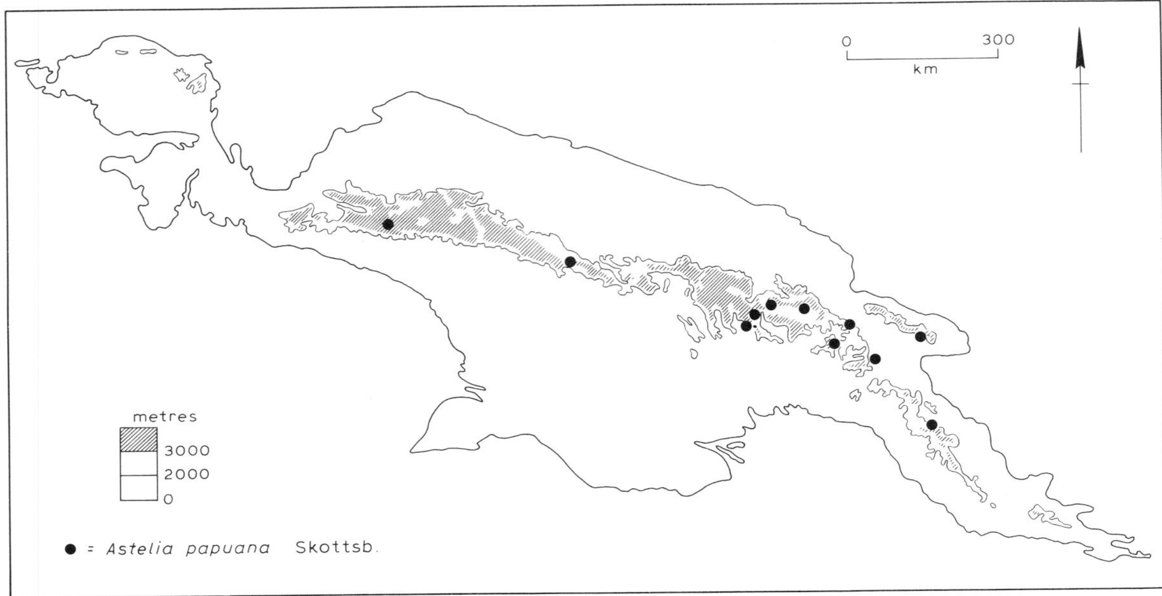

Fig. 5.2. Distribution of *Astelia papuana*. (Based on collections in Herbarium Australiense.)

The beauty of the above model is that, as shown in Fig. 5.3, it will accommodate all three of the environmental changes for which we are searching; reduction in sea levels, cooler climate and increased aridity. However, a model is all very well, but what is the empirical evidence that such events actually occurred in the tropics, and specifically in the Malay archipelago?

SEA-LEVEL CHANGES

Quaternary

The frequency with which sea levels have varied through the Quaternary is well illustrated by oxygen isotope data (Shackleton and Opdyke 1973, 1976) because oxygen 16:18 variations mainly reflect the volume of water tied up in glaciers (Shackleton 1967) and thus the amount extracted from oceans. For the last Glacial period sea levels can also be determined, with considerable precision, from the occurrence of subaerial geomorphological and other features on the sea bed. The maximum lowering of sea level in the Quaternary was about 200 m, which would have exposed the Sunda platform and the Sahul shelf. Batchelor (1979) has argued that, for the Malesian region, sea levels reached their minimum, and hence land areas were most extensively exposed, during the middle Pleistocene. There is other evidence too that these shelves were above sea level. Submarine channels on both shelves show the probable courses of rivers across them (Verstappen 1975). Submarine peat deposits have been described from the Sunda shelf (Aleva 1973; Biswas 1973). On the Sahul shelf, kunkar nodules have been found at depth (van Andel *et al.* 1967). Finally, in the Gulf of Carpentaria, there is evidence for old coast lines, subaerial erosion, and a former freshwater or brackish-water lake at a depth of *c.* −60 m (Torgersen *et al.* 1985).

Sea levels slightly above present ones are recorded during Holocene time (Tjia *et al.* 1984).

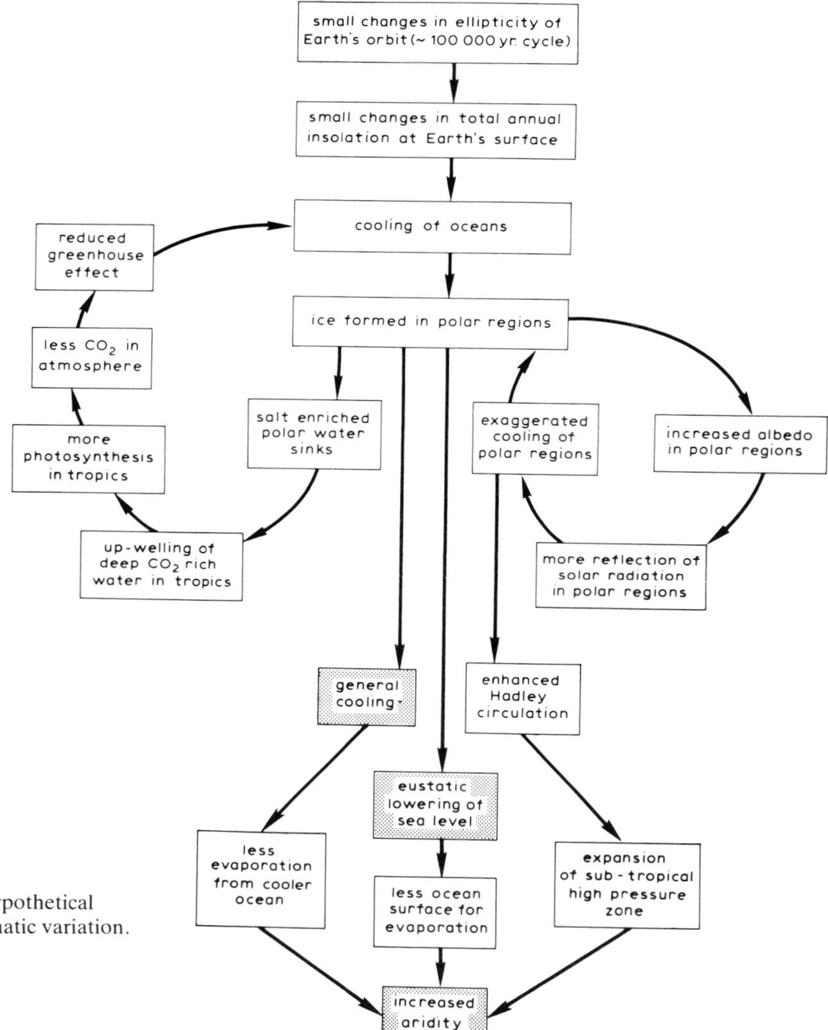

Fig. 5.3. A generalized hypothetical model for Quaternary climatic variation.

Late Tertiary

Sea-level changes during the Tertiary have been widely reported (Vail, Mitchum and Thomson 1977), although these may have resulted more from tectonic than astronomical causes. The main changes are probably due to both local and world-wide tectonic events; perhaps the overriding factor for the Tertiary was the effect of the expansion of the oceans due to sea-floor spreading, resulting in marked changes in ocean volumes. The mid-Oligocene onset of Antarctic glaciation also caused a marked lowering of global sea levels (Kennett *et al.* 1975). Local tectonic events are superimposed on world-wide changes, and have resulted in many complex sea-level histories for different parts of the Malay archipelago.

The degree to which the distribution of land and sea in the region has altered during the late Cainozoic can be illustrated by comparing an early Miocene reconstruction (Fig. 5.4), with the present-day and Pleistocene reconstruction presented in Fig. 5.5.

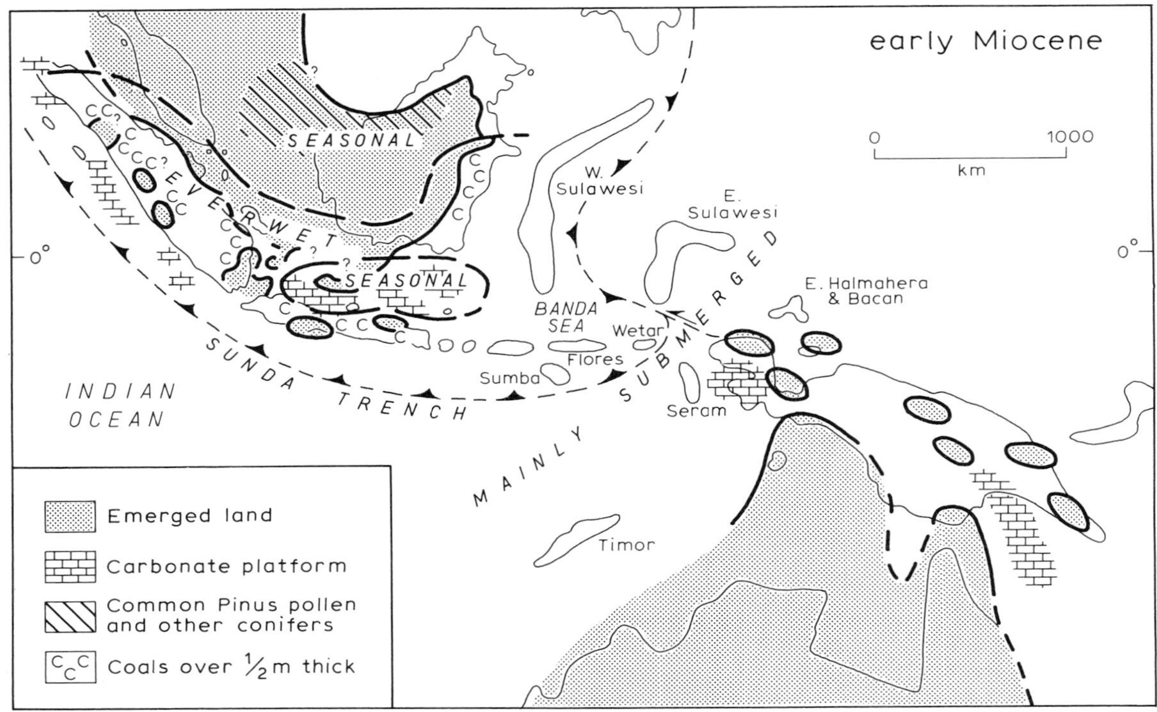

Fig. 5.4. Early Miocene palaeogeographical reconstruction of the Sunda–Sahul region. (Position of Asian and Australian plates, and distribution of ocean floor according to M. C. Audley-Charles (unpublished), coastlines after Doutch (1972), Paltrinieri and Saint-Marc (1976), and unpublished data.)

CHANGING SEASONALITY

There have also been past changes in total rainfall, seasonal distribution of rainfall and changes in the precipitation/evaporation ratio.

Increasing evidence is now emerging from various sources for more strongly seasonal climates within both the late Tertiary and the Quaternary. Also mention has frequently been made of an hypothetical 'savanna corridor' (Ashton 1972), along which Pleistocene mammals and other organisms could have migrated from Thailand to Java and Sulawesi, where extensive mammalian faunas of Asiatic origin have been discovered (Sartono 1973).

Quaternary

The best record for greater climatic seasonality in the Quaternary is from the Atherton Tableland 17°S in Queensland. Here Kershaw (1976) has produced pollen evidence for a striking vegetational change. During the last glaciation the area apparently bore sclerophyll vegetation, whereas during both the last and the present interglacial, rain forest has predominated. The reduction in total rainfall suggested is of the order of 60 per cent. Much of this could, however, result from the coastline being further away because of eustatic lowering of sea level.

From the Sahul shelf we have the kunkar nodules mentioned above, which form only under arid climates. Their presence indicates a strongly seasonal climate at a time of lowered sea level (i.e. presumably glacial). Morley (1982) showed that the climate of south Kalimantan was of a much more seasonal character in the early Holocene, although he had concluded (1981) that there was no evidence for a formerly drier or

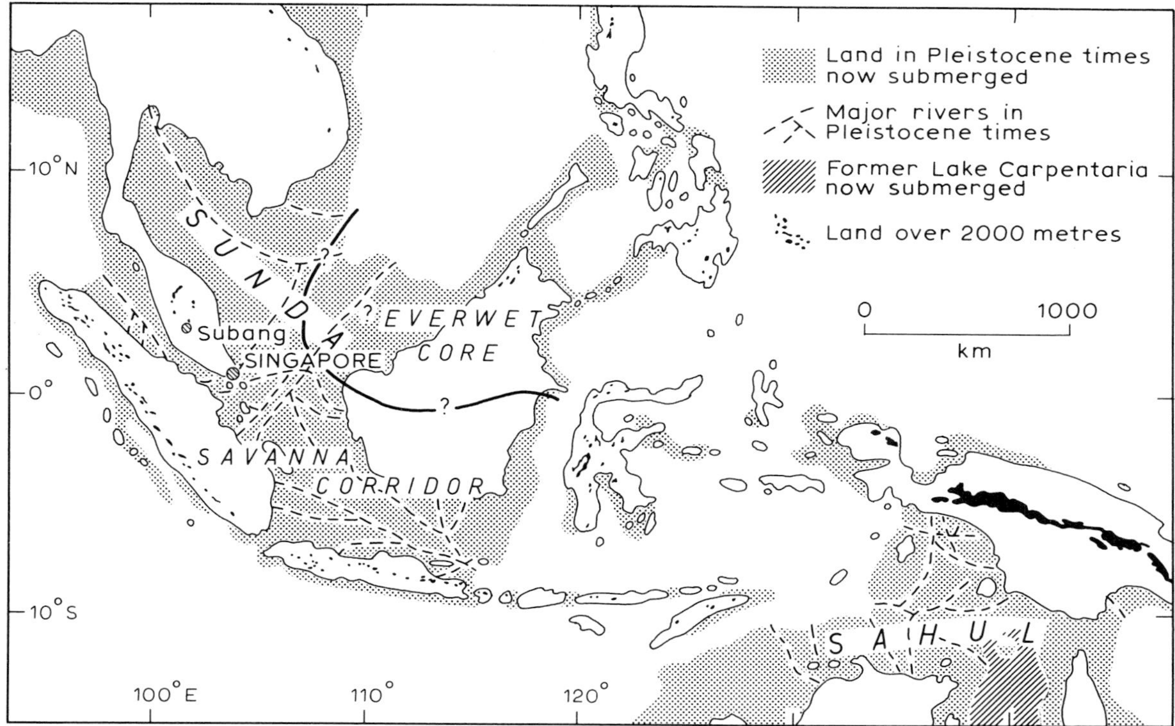

Fig. 5.5. A generalized palaeogeographical reconstruction of the Sunda–Sahul region during one of the many Quaternary glacial maxima. A possible interpretation of the 'savanna' corridor is suggested for the Sundaland middle Pleistocene (see text). Ever-wet refuges may also have occurred in montane areas.

more seasonal climate within the Holocene and latest Pleistocene in central Sumatra.

Some interesting new evidence for a period of seasonal climate is now available from the Malay peninsula. A peat/detritus mud sample from a deposit considered to be of middle Pleistocene age (B. C. Batchelor, personal communication) from Subang near Kuala Lumpur yielded the pollen assemblage shown in Table 5.1. This remarkable assemblage is dominated firstly by *Pinus* pollen and secondly by that of grasses. The sample was one of several collected from a stratigraphical sequence transitional between the Old and New Alluvium, from an active tin mine. Unfortunately, when the significance of the sample was appreciated, a re-examination of the sampling locality failed to yield, despite extensive searching, a similar section and hence a detailed sequence of samples through the peat was never obtained. Peats are common in this transitional sequence in the Kuala Lumpur region. Many have been examined for palynology, and all others (though not yet published) so far yield assemblages which can be interpreted in terms of typical ombrogenous peat swamp or alluvial swamp settings. The peat sequence from which the sample in Table 5.1 came lay stratigraphically below all of the others collected so far and yielded none of the pollen types characteristic of these swamp types. The combination of abundant *Pinus* and Gramineae pollen is suggestive of the pine woodlands of Thailand (Phengklai 1972) and Luzon (Kowal 1966) although it is well known that these communities are considerably encouraged by Man's activities, especially burning. The assemblage clearly indicates a more markedly seasonal climate than that of present day Malaya (where *Pinus* does not occur) and by inference it can be suggested that the mid-Pleistocene ever-wet rain forests were pushed much further south. Other notable features of this pollen assemblage are the absence of Malesian rain forest pollen types, and the presence of *Dacrydium* pollen. *Dacrydium* is rarely found in lowland settings in Malaysia outside peat swamps (Morley 1981) or heath forest (Brunig 1974), but it occurs commonly in streamside communities in Thailand, often associated with rattans. *Calamus* pollen was also a common constituent of this sample and may have been derived from rattan brakes (*sensu* Stamp and Lord 1923) typical of seasonal streamside settings in Indo-China.

Table 5.1

Palynomorphs recovered from sample 3/15, from Subang, near Kuala Lumpur, Malaysia. Age: middle Pleistocene

Pinus	34%
Gramineae	18%
Smooth Fern Spores	18%
Lycopodium cernuum	7%
Cyathea	7%
Cyperaceae	2%
Pandanus	1%
Ilex	2%
Calamus	2%
Compositae	1%
Pteridium	1%
Nauclea type	Tr
Myrtaceae	Tr
Palaquium type	Tr
Stenochlaena palustris	Tr
Altingia excelsa	Tr
Calophyllum	Tr
Dacrydium	Tr
Gleichenia	Tr
Dillenia type	Tr
Shorea type	Tr
+ 42 other types	
Other palynomorphs:	
Concentricystes circulus	2%
Gymnosperm tracheids (?*Pinus*)	Tr

The character of this assemblage provides the first real evidence for a savanna corridor in Malaya through which savanna plants and animals could have migrated in the mid-Pleistocene.

Verstappen (1975) argued that many Sumatran landforms could have formed only under a 'semi-arid' climate. Such conditions are also suggested by geologists to account for diagenetic features of the Old Alluvium of peninsular Malaysia and Singapore.

A possible reconstruction of middle Pleistocene climates is given in Fig. 5.5.

Late Tertiary

New lines of evidence are emerging from which it is possible to infer seasonality of climate for the late Tertiary. One source of palaeoclimatic data comes from the distribution of extensive late Tertiary carbonate platform sequences. Such massive platforms (analogous to the Australian Great Barrier Reef) accumulate today off coasts where the climate is, in general, markedly seasonal and where drainage water therefore tends to carry less fine sediment than in ever-wet climates. Secondly, ever-wet climates may be inferred from the distribution of coal seams (over 0.5 m in thickness), implying peat swamp conditions which occur only under an ever-wet climate (Morley 1982). Another line of evidence comes from the pollen record in Brunei where Muller (1966) recorded common gymnosperm pollen, especially *Pinus*, together with *Abies*, *Tsuga*, *Keteleria* and *Ephedra*, which are all taxa that occur today under seasonal climates in Indo-China. This assemblage has subsequently been recorded from much of the South China Sea region (Morley 1978), and it is considered highly likely that the assemblage also implies a strongly seasonal palaeoclimate. Fig. 5.4, showing land areas for the early Miocene, also includes a reconstruction of inferred climate based on the above criteria.

A more seasonal climate may also be indicated for the late Miocene and Pliocene of Brunei (Muller 1972) and the Natuna Sea (Morley 1978) from the occurrence of *Aegialites* pollen, a genus which, as we have already noted (Fig. 5.1), today exhibits a markedly disjunct distribution.*

*Further evidence of past seasonal climates is provided by several strata of grass pollen of Plio-Pleistocene age from a well drilled at Misedor in the Mahakam delta. (Caratani, C. and Tissot, C. (1985). *Etudes de Geographie*, CEGET, Universitaire de Bordeaux.)

THERMAL CHANGE

The evidence for thermal change comes from four main sources. These are oxygen isotope and foraminiferal evidence from the oceans, geomorphological evidence for lowered snowlines, and palynological evidence for changes in altitudinal zonation of vegetation.

Quaternary

The oxygen isotope evidence from cores obtained below tropical oceans is important because ocean cores can give continuous records throughout the Quaternary. For instance core V28–238 from the western equatorial Pacific showed a sequence of at least 21 oscillations relating to palaeoglaciation, and hence indirectly to fluctuations in sea surface temperature, over the last 1.6 Ma (Shackleton and Opdyke 1973, 1976). However, integration of oxygen isotope and foraminiferal data over the world ocean surface suggested a lowering of ocean surface temperature of only 2 °C at the peak of the last Ice Age, *c.* 18 000 years BP, in the region of the Malay archipelago (CLIMAP 1976), and a later integration (CLIMAP 1981) gave sea temperatures which, in most areas, are even nearer to present-day ones.

Only three islands within the Malay archipelago are known to have been glaciated. The New Guinea mountains still bear permanent snow, and glacial lakes and other features suggest that the snowline could once have been *c.* 1000 m lower than at present (Reiner 1960; Hope and Peterson 1975, 1976; Löffler 1984). Evidence of glaciation on Mt. Kinabalu in Borneo was described by Koopmans and Stauffer (1968). Glaciers formerly extended down to *c.* 3000 m, although glacial debris was carried by streams or gravity to much lower altitudes. Glacial evidence from Gunung Leuser and other mountains in north Sumatra has now been described (van Beek 1982; Whitten *et al.* 1984).

Minimum ages for the last deglaciation of most of these areas are reasonably concordant (Flenley and Morley 1978) and are mostly in the range 14 000 to 9000 BP depending on altitude. However, the age of 7590 ± 40 BP for deglaciation near the summit of Gunung Kemiri (3340 m) in Sumatra is very late for this altitude. If this area was formerly glaciated (van Beek 1982), the lake deposit providing this age probably dates from well after the melting of any former glaciers.

Palynological evidence for thermal change comes from New Guinea and Sumatra/Java. From New Guinea we have evidence, principally from above 1900 m, for vertical movement of the forest limit (Fig. 5.6). It seems clear that in the late Pleistocene the forest limit was as much as 1700 m lower than it is today. Löffler (1984) has argued that this is an over-estimate, since the pollen sites used could have been frost pockets. This could be true for some sites, but is unlikely to be true for all, and especially not for the Sirunki swamp, the best dated and most significant site (Walker and Flenley 1979), which is on a large open watershed.

From Sumatra and west Java we now have evidence from lower altitudes (900–1600 m) for changing altitudinal boundaries within the forest (Fig. 5.6). The principal boundary we have used is one within the lower montane forest (*sensu* Whitmore 1975), between a lower zone of forests (montane forest I of Morley 1982) which tend to be dominated by *Castanopsis*, *Quercus* and Lauraceae, and an upper zone (montane forest II of Morley 1982) which includes the conifer *Dacrycarpus* (*Podocarpus* s.l.) as an important tree, together with *Engelhardia* and *Eugenia*. This boundary appears to have been at least 400 m lower during the late Pleistocene, and possibly 600–800 m lower (Morley 1982; Maloney 1980, 1981; Stuijts 1984; Flenley 1984, 1985).

It will be clear from the above that the estimates of Pleistocene altitudinal shifts are widely disparate. An attempt to combine them into a unifying hypothesis (Fig. 5.7) has been made by Walker and Flenley (1979). If the present temperature lapse rate of 0.6 °C per 100 m were increased in glacial times to 0.8 °C, the former altitudes are mutually conformable, with one exception (Fig. 5.7)—the firn line seems

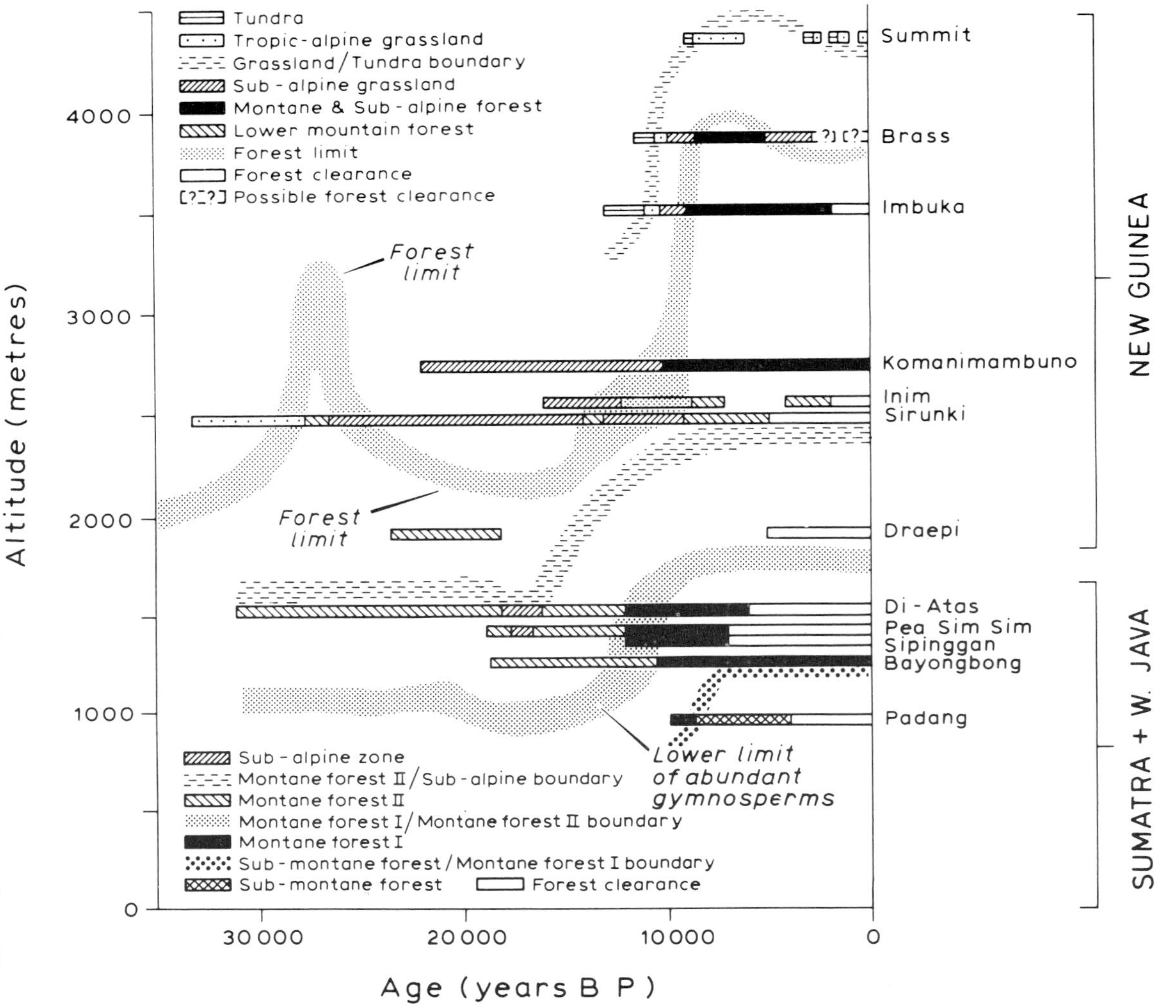

Fig. 5.6. Late Quaternary vegetational changes in the New Guinea highlands, Sumatra, and West Java, as judged from palynological evidence. (After Flenley 1985.)

too high. A steeper lapse rate, however, implies drier air and would be compatible with reduced precipitation. Perhaps, therefore, the firn line was artificially high because of limited precipitation. In the lowlands the reduction in precipitation implied by this hypothesis could have resulted in an increase in the extent of seasonal (monsoon) forests, and concomitant reduction in the extent of rain forests. This could have permitted the migration of seasonal-climate plants and animals discussed earlier (Figs 5.1, 5.5).

Late Tertiary

For the late Tertiary, there is less satisfactory evidence for thermal change. This is due largely to the fact that although for the late Quaternary, palynological sites can be found in upland areas which have ideal characters for the recognition of

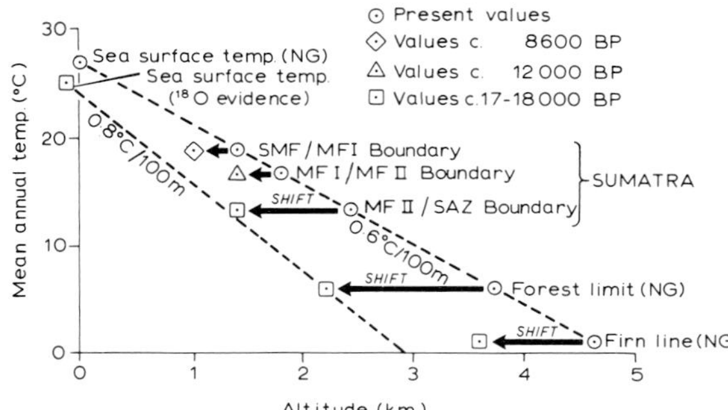

Fig. 5.7. The relationship between mean annual temperature and altitude in New Guinea and Sumatra at present and in the late Quaternary 17 000–18 000 BP, the latter as judged from palynological evidence. The data suggest the temperature lapse rate (dashed line) may have been higher in the past (SMF: submontane forest; MF I, MF II: the two zones of lower montane forest; SAZ: subalpine zone). (Based on Flenley 1984.)

vertical movements of vegetation zones, and from which inferences of thermal climate changes can be made, all such pre-Quaternary sites would have been lost through erosion, unless tectonically controlled intra-montane basins, active over millions rather than thousands of years, can be found. Such examples do occur elsewhere, and have been studied palynologically. For instance, a very long pollen record is available from the Sabana de Bogotà (Colombia). The remarkable pollen record recently published from that area (Hooghiemstra 1984) shows no fewer than 27 complete oscillations in the last 3.5 Ma and demonstrates the amazing mobility of tropical mountain vegetation. No such basins have yet been satisfactorily identified in Malesia although deep drilling may confirm such sequences in Java or the Barisan range of Sumatra.

However, some evidence for Tertiary temperature changes is available from foraminiferal data. Van Gorsel and Troelstra (1981) demonstrated changes in the ratio of cold and warm water foraminifera from the Javan Mio-Pliocene, and show that there is evidence for changes in the ratio of cold and warm water foraminifera in the Messinian (late Miocene). This cold phase is thought to have analogues in the Messinian of the Mediterranean and the late Miocene of New Zealand (Kennett and Watkins 1974), and hence is considered to relate to world-wide cooling.

CONCLUSIONS

During the Pleistocene, glacial periods occupied much more of the total time than interglacials, and for the greater part of the last two million years, therefore, the Malay archipelago may have looked roughly as shown in Fig. 5.5. During glacial periods land connections were increased, mountain vegetation had many more stepping-stones on which to migrate, and there was clearly a savanna corridor allowing migration of plants and animals from Thailand to Java and Sulawesi.

The evidence from the Tertiary suggests that the pattern of land, sea, and climate showed marked differences across the area compared with the present day and that there must have been migration of taxa across the region well before the beginning of the Quaternary era. Clearly, further reconstruction of both Quaternary and late Tertiary palaeogeography of the area, and further analysis of data providing evidence for changing seasonality of climate, will allow a much better understanding of present and former plant distributions and could well permit major disjunctions and migrations to be dated with considerable accuracy.

ACKNOWLEDGEMENTS

We are particularly grateful to Mrs J. C. Newsome for allowing us to use her unpublished data, and to Mr K. Scurr for drawing the figures. RJM is grateful to the directors of Robertson Research International for permission to publish.

6 BICENTRIC DISTRIBUTION IN MALESIA AS EXEMPLIFIED BY PALMS

John Dransfield

*In Malesia we may recognize three main distribution patterns of plants: those which are widespread, those which are centred either in the west (Sundaland) or the east (Papuasia), and those which are bicentric, with concentrations in both west and east yet absent from the central region. Analysis of distribution patterns rests heavily on good taxonomy and the palms are highly suitable: there is a new generic monograph and they are an ancient group, one of the first modern families of monocotyledons recognizable in the fossil record. The palm family is pan-tropical and has about half its genera and species (*c.*97,* c.*1400 respectively) in the eastern tropics.*

Detailed analysis of the eastern members of the five subfamilies represented strongly suggests there are two distinct elements in Malesia, of northern and western Laurasian–Tethyan, and southern and eastern Gondwanic origin respectively.

The most primitive of all palms occur in the northern hemisphere. Analysis in the light of current taxonomic understanding suggests the family evolved in Pangaea, and then northern and southern groups evolved separately, in Laurasia and Gondwanaland. By the late Oligocene the two groups were both very diverse and contained several modern genera. Sundaland contained Laurasian taxa plus others arrived from Gondwanaland, either carried northwards on India or arriving via the Gondwanic island fragments which had rafted from Australia–New Guinea (Chapter 2). The Oligocene Papuasian palms were, however, all Gondwanic. By late Miocene the Sundaic and Papuasian palm floras had been brought into close contact and since then have intermingled to various extents. Thus originated the remarkable bicentricity which the family shows in Malesia, more strongly developed than in any other angiosperm family. This analysis accounts for most Malesian palms, leaving a few serious problems (e.g. Metroxylon, Pigafetta*) to await further phylogenetic analyses.*

INTRODUCTION

It needs to be stressed from the very beginning that an analysis of the distribution patterns of plants in the Malesian region depends very heavily on there being a good taxonomic understanding of the taxa involved. This is of course a truism. It will be emphasized later when we come on to discuss the distribution of the palms in detail.

The making of modern Malesia by the final juxtaposition of the Sunda and Sahul shelves in the mid-Miocene seems now to be well established (Audley-Charles *et al.* 1981, and this volume, Chapter 2). The geological history of the region is often rather precisely reflected in the geographical ranges of taxa, and there is now a considerable body of literature dealing with the relationships of geology with distribution in the region (e.g. Schuster 1972, 1976; Raven and Axelrod (1972, 1974); Johnson and Briggs (1975); Whitmore (1981)).

DISTRIBUTION TYPES

In the Malesian region we can distinguish several major patterns of distribution which we can divide broadly into three main types: taxa which are widespread throughout the region, taxa which are present only at the Sunda or Sahul ends of Malesia, and taxa which are present both on the Sunda shelf and in Papuasia but which are absent from the central area of Sulawesi and much of the Moluccas.

Widespread taxa

The genus *Rhododendron* provides an excellent example of a taxon widespread throughout the region. Although the number of species in the centre of Malesia is not great, the number in New Guinea and in Sundaland is very great indeed. Only one species, however, reaches Australia and one other the Solomon archipelago. Thus, although the genus is widespread, the overwhelming abundance of species in the northern hemisphere, e.g. in the Himalaya, gives a northern bias to this genus. Nevertheless it is widely dispersed within Malesia.

At the family level Dipterocarpaceae are also widely dispersed throughout the region but when viewed in the context of the distribution of the family both outside and within the region, a strong north-western bias becomes obvious; both at species and generic level, there is very much greater diversity west of Wallace's line (Fig. 8.3 in Whitmore 1981).

The genus *Corybas* (Orchidaceae) is also widespread throughout the region, occurring on mountain tops and displaying quite remarkable diversity in New Guinea. It occurs from southern China and the Himalaya through to New Zealand and the sub-Antarctic islands. Although the genus is so widespread there is a fairly clear austral bias in terms of species number and morphological diversity (van Royen 1983; Dransfield, Comber and Smith, 1986).

As a final example of a widespread taxon, the genus *Styphelia* (Epacridaceae) occurs throughout the archipelago in montane forests and lowland heath forest. However the genus and family display a strong austral bias. The distribution of *Styphelia* is paralleled by other genera adapted to similar poor soil conditions, e.g. *Leptospermum* and *Baeckia* in the Myrtaceae.

In these four examples, although the taxa are widespread, bias in terms of numbers of species and the distribution outside the Malesian region strongly suggests the dispersal of the taxa across Wallace's line, even though the taxa have not been submitted to a careful biogeographical analysis.

Sahul and Sunda distribution

Other groups by not transgressing Wallace's line show an even stronger Sahul or Sunda bias. Thus the genus *Quercus*, widespread in the northern hemisphere, is well represented in Sundaland yet not one species crosses Wallace's line. The genus *Grevillea* in the Proteaceae displays just the opposite; it is represented by about 190 species in Australia, New Caledonia and Vanuatu, with four species in the Malesian region in Sulawesi, the Moluccas and New Guinea. Superficial analysis of both *Quercus* and *Grevillea* suggests origins in Laurasia and on the Australian plate respectively; the two genera have not been successful in crossing Wallace's line.

Disjunct distributions

In the third present-day distribution type, a taxon is represented both in Sundaland and on the Sahul shelf but is absent from the central area of the Malesian region. The high altitude umbellifer, *Oreomyrrhis andicola*, for example, is present in Malesia on Mt. Kinabalu, Borneo and in the mountains of New Guinea; elsewhere it is widespread in the Andes. In this instance we may postulate that its distribution is disjunct because of the absence of suitably high mountains between Borneo and New Guinea. Its distribution is also suggestive of efficient long-distance dispersal.

We may postulate, however, that if a taxon were sufficiently ancient, it might possibly have arrived in the Malesian region from two different directions, one from the north, the other from the south. The most famous and most highly analysed example of such a distribution is surely the Fagaceae. This family is divided into three subfamilies of which the Quercoideae and the Castanoideae are well developed in west Malesia. On the other hand the third subfamily, Fagoideae, is represented by *Fagus* in the northern hemisphere and *Nothofagus* in New Guinea,

Australia, New Caledonia, Tasmania, New Zealand, and South America. Whatever the origin and dispersal routes of the Fagaceae may be, and their interpretation is certainly very controversial (van Steenis 1971; Schuster 1972; Whitmore 1981; Humphries 1981), it does seem reasonable to suppose that Malesian representation of the family is derived from two sources, mainland Asia and austral.

MALESIAN PALMS

From these few examples it must be obvious that plant distributions in the Malesian region are very varied; species and genera may, of course, be much less widespread than families. In this paper the origin and relationships of the Malesian representation of one large tropical family of flowering plants, the palms, will be examined in some detail. I have already analysed the Malesian palm flora in a preliminary fashion (Dransfield 1981), but I believe the present paper has a much firmer basis and allows me the opportunity to correct and alter speculative ideas presented in the first paper.

The palms are particularly interesting from a phytogeographical viewpoint. They are known to be ancient, being one of the first recognizable modern families of monocotyledons to be present in the fossil record, occurring first in the Upper Cretaceous (Daghlian 1981; Muller 1981). However, in view of the remarkable parallels in pollen structure as viewed in electron microscopy in unrelated extant palms (Ferguson, in press), the whole fossil pollen record of palms needs critical reappraisal. The family is found throughout the tropics and subtropics, but displays a relatively high degree of endemism, perhaps related to the fact that the propagules are generally rather large. We may suggest palms more nearly reflect past geography than do efficient, wind-dispersed taxa such as *Rhododendron*. Finally, the family is now quite well understood at the generic level, a major new classification of the palms having just been completed (Uhl and Dransfield, in press). Although Natalie Uhl of Cornell University and I have only just begun a cladistic analysis of the family, input from a preliminary analysis and from a much better understanding of subtribal and tribal relationships has produced a basis for a reassessment of the Malesian palms and their relationships.

The palms of the eastern tropics including India, China, continental south-east Asia, Malesia, Australia, and the west Pacific number about 1400 species belonging to about 97 genera (Moore 1973*a*). Thus just under half of all palm genera and half of all known palm species are found in this region. The great richness in terms of genera and species is almost certainly associated with the ancient history of the family coupled with the complex geological history of the region.

Moore (1973*b*) hypothesized an origin of the family in the Cretaceous in west Gondwanaland followed by great diversification and dispersal during the early Tertiary into Laurasia, coupled with considerable rafting on fragments of Gondwanaland.

In what follows I shall analyse the palm flora of Malesia in terms of subfamilies, tribes, and subtribes and attempt to show the bicentric nature of the family before finally suggesting a narrative for origin and dispersal.

We (Uhl and Dransfield, in press) have divided the family into six subfamilies: Coryphoideae, Calamoideae, Nypoideae, Pseudophoenicoideae, Arecoideae, and Phytelephantoideae. Of these the Phytelephantoideae, is confined to South America; it is an extremely peculiar, isolated and specialized group which will not be discussed further in this paper.

Subfamily Coryphoideae consists of three tribes, all of which have representatives in the Malesian region.

(1) Tribe Corypheae are divided into four subtribes.

(a) Subtribe Thrinacinae comprise some of the least specialized members of the subfamily, an assessment based on important recent work on major trends of evolution in the family by Moore and Uhl (1982). Thrinacinae have a very interesting distribution, occurring in South America (4 genera), the Caribbean (3 genera), North America (1 genus), Europe (1 genus), and in east Asia (4 genera); one, *Maxburretia* (Fig. 6.1), occurs within Malesia in Malaya. These are all apocarpic palms.

Fig. 6.1. *Maxburretia rupicola* (Coryphoideae, Corypheae, Thrinacinae) a primitive apocarpous fan palm endemic to limestone hills near Kuala Lumpur, Malaysia.

(b) Subtribe Livistoninae on the other hand are syncarpic. The subtribe has a strong northern hemisphere bias. Of the 12 genera, one genus in the Americas (*Copernicia*) also occurs in the southern hemisphere. In south-east Asia, *Livistona*, *Pholidocarpus*, and *Licuala* transgress Wallace's line, while *Pritchardiopsis* (endemic and on the verge of extinction in New Caledonia) and *Pritchardia* (in Hawaii, Fiji, and Samoa) are Pacific in distribution. The distribution of the whole subtribe together with a rich fossil record suggests its early differentiation in Laurasia followed by later dispersal into the southern hemisphere. *Livistona* and *Licuala* are of particular interest. *Licuala* seems almost equally richly represented at the two extremes of Malesia but when viewed in the context of its closest relative, *Johannesteijsmannia* (which is endemic to Sundaland), and *Pholidocarpus* (predominantly in Sundaland), it seems more reasonable to suppose that the Papuasian representation is the result of dispersal from Sundaland. *Livistona*, on the other hand, seems more richly represented in Australia than at the western end of Malesia. This is the only member of the subtribe, apart from one species of *Licuala*, to be represented in Australia. Outside Malesia and Indo-China the genus is also present in the Horn of Africa and Arabia (*Livistona carinensis*—see Dransfield and Uhl 1983). I earlier postulated that *Livistona* may have entered Malesia from two ends (Dransfield 1981). However, the presence of only a few widespread taxa in South America reinforces the Laurasian bias and suggests to me that the Australian *Livistona* species may have originated in invasion from Sundaland; fragmentation of the ranges of *Livistona* in Australia due to climatic deterioration may have led to speciation. Far from being a Gondwanic relic, *Livistona* may, I now believe, be an immigrant.

(c) Subtribe Corypheae also have a strong Asiatic bias. Of the four genera, all are in Asia, but only one, *Corypha*, is in Malesia; it is represented by one species, *C. utan* (Fig. 6.2). *Corypha utan* is very widespread and occurs more or less as a pioneer weed in coastal storm forest throughout Malesia to Australia. Its presence east of Wallace's line seems to be the result of relatively recent dispersal.

(d) Subtribe Sabalinae, the fourth subtribe, is found only in the Americas.

(2) Tribe Phoeniceae comprise the single genus *Phoenix*, another relatively unspecialized palm with an apocarpic gynoecium. It is distributed

Fig. 6.2. *Corypha unbraculifera* (Coryphoideae, Corypheae, Coryphinae), the Talipot Palm, native to southern India. Here in flower in the Lake Gardens, Taiping, Malaysia.

Fig. 6.3. *Phoenix paludosa* (Coryphoideae, Phoeniceae), the Mangrove Date Palm, gregarious in mangrove forests of Sundaland. (Perak, Malaysia.)

throughout Africa (one species), the Mediterranean, Arabia, and India to the Far East. One species (*Ph. paludosa*, Fig. 6.3) occurs in mangrove forest in Sundaland. There seems to be a strong Laurasian bias of the tribe, despite the abundance of one species in Africa.

(3) The Borasseae, third tribe of the Coryphoideae, appear to have had a quite different history. There are seven genera distributed in the land adjoining the Indian Ocean—three in Africa, two in Madagascar, one in the Seychelles, one in the Mascarenes, two in India, and two in the Malesian region. *Borassodendron* comprises two species, one in Malaya (Fig. 6.4) and one in Borneo (Dransfield 1972); one widespread cultigen of *Borassus* occurs throughout the region and another, *B. heineana*, remains very poorly known, occurring in New Guinea. The present-day distribution of the subtribe suggests an origin in Gondwanaland followed by rafting on the Indian plate; the Double Coconut (*Lodoicea*) survives in the Seychelles, and *Borassodendron* and *Borassus* may have reached their present position by dispersal after the collision of India with Laurasia. An alternative route may have been provided by rafting on the Gondwanic fragments of Australia destined to become parts of the Sunda shelf (Audley-Charles, this volume, Chapter 2).

Subfamily Calamoideae (Lepidocaryoid Major Group of Moore 1973*a*) consists of twenty-two genera and is the largest in terms of species in Malesia. We recognize two tribes: (1) tribe Calameae (with pinnate or pinnately nerved leaves) and (2) tribe Lepidocaryeae (with palmate leaves). The latter tribe contains three genera confined to the New World. Of the Calameae four genera are confined to Africa (except for one species of *Raphia* in South America). The remaining 15 genera are concentrated in Malesia, where all but two show a strong western bias; indeed the distribution of the Asian Calamoid genera appears like a series of concentric circles centred on the Malay Peninsula

Fig. 6.4. *Borassodendron machadonis* (Corypheae, Borasseae), a Malayan endemic. (Waterfall Gardens, Penang.)

Fig. 6.5. *Metroxylon salomonense* (Calamoideae, Calameae) the Solomon's Sago Palm, with its compound terminal inflorescence and most leaves removed for thatch. *Cocos nucifera* (Arecoideae, Cocoeae) the coconut. (Honiara, Solomon Islands.)

(Fig. 6.7 in Dransfield 1981). The further away from this centre the fewer the genera and the fewer the species. However, this very tribe displays marked bicentricity. Two genera, *Pigafetta* and *Metroxylon* (Fig. 6.5), are, in their natural state, confined to east of Wallace's line. The least specialized members of the tribe in terms of floral structure and arrangement are two African genera, *Eremospatha* and *Laccosperma*. At present the diversity in south-east Asia seems best explained in terms of origin in the late Cretaceous or early Tertiary in Africa followed by dispersal along Tethys to Sundaland. There are fossil pollen records referred to the Calamoid genus *Eugeissona* in the Oligocene of Borneo (Muller 1968) which at least suggests the presence of the tribe in Sundaland in the early Tertiary. Papuasian *Metroxylon* is included with Sundaic *Korthalsia* in the same subtribe based on the peculiar structure of the flower-bearing branches and floral structure. However, in terms of habit, the two genera are very different, *Korthalsia* being a rattan (climber) and *Metroxylon* a massive tree palm. *Pigafetta* is morphologically very isolated, although clearly a member of the tribe. *Metroxylon* occurs in New Guinea, the Solomons, Vanuatu, Fiji, Samoa, and the Carolines, while *Pigafetta* is confined to Sulawesi, Maluku, and New Guinea. The distribution of these two genera and the three remarkable fan-leaved genera (tribe Lepidocaryeae) in the Americas are at present extremely difficult to account for. It seems improbable that *Pigafetta* and *Metroxylon* reached their present position by late-Tertiary dispersal across Wallace's line. Perhaps they reached their present ranges by much earlier dispersal across the fragments of Gondwanaland in the Malesian region, as described in Chapter 2. Results of a cladistic analysis of the subfamily now in progress may clarify relationships and allow the development of a phytogeographic narrative. I do not rule out a trans-Pacific link with Tribe Lepidocaryeae.

Subfamily Nypoideae contains the single monotypic genus *Nypa* (Fig. 6.6). Although morphologically isolated *Nypa* retains several unspecialized features such as the apocarpic gynoecium with ascidiform carpels (Uhl 1972). At present

Fig. 6.6. Fruit head of *Nypa fruticans* (Nypoideae), found in mangrove forests from the Bay of Bengal to the Pacific but with fossils from Upper Cretaceous strata in all tropical continents and in the Eocene London Clay of southern England. (Selangor, Malaysia.)

the palm occurs in mangrove forests from the Bay of Bengal through to the west Pacific. The pollen is distinctive and easily recognized, which has no doubt contributed to its long fossil record back to a 'more or less simultaneous appearance' in the Maastrichtian (Upper Cretaceous) of Borneo, India, Africa, and tropical South America (Muller 1981). It was formerly very much more widespread; it occurred, for example, in Europe and Australia during the Eocene, and it appears to have been an important mangrove plant of the Tethys Ocean. Its clear presence in Sundaland and Australia before the Miocene, if correctly based, suggests that it successfully colonized the postulated island chain between Sundaland and the Sahul shelf. *Nypa* fruits are adapted to dispersal by floating in sea-water so rapid dispersal seems very likely.

Subfamily Ceroxyloideae is divided into three tribes but we shall consider only TRIBE CEROXYLEAE, the others being confined to the New World and the Mascarenes. One member of tribe Ceroxyleae occurs in Queensland. The well-known Australian palm *Orania appendiculata* (Fig. 6.7) has recently been shown to have no relationship with the genus *Orania* (see below) but to represent a new genus, *Oraniopsis*, related to the Andean wax palms, *Ceroxylon* (Dransfield, Irvine and Uhl 1985). Tribe Ceroxyleae has a classical Gondwanic distribution with one genus *Ceroxylon* in the Andes of South America, two genera *Ravenea* and *Louvelia* in Madagascar, *Juania* on Juan Fernandez island, and *Oraniopsis* in Queensland. However, the very close relationship between *Oraniopsis* and *Ceroxylon* has suggested to us that *Oraniopsis* may have reached Australia by dispersal via a south Pacific route rather than by rafting. Palm pollen in Tertiary sediments in Antarctica (Cranwell *et al.* 1960) suggests that such a route may not have been impossible. In many ways *Oraniopsis* and *Ceroxylon* parallel the distribution of *Nothofagus*.

We now come to the complex **subfamily Arecoideae**. This taxon includes all palms in which the flowers are borne in triads or triad-derivatives in the inflorescence; a triad is a flower group consisting of a central pistillate and two lateral staminate flowers. There are six tribes. Tribes IRIARTEEAE (1) and GEONOMEAE (2) are confined to the New World and PODOCOCCEAE (3) to West Africa so will not concern us further. Tribe CARYOTEAE (4) is of great interest. The three genera *Arenga* (Fig. 6.8), *Caryota* (Fig. 6.9) and *Wallichia* possess many unusual features, such as induplicate vernation of the leaves and the frequent presence of basipetal, hapaxanthic

Fig. 6.7. *Oraniopsis appendiculata* (Ceroxyloideae, Ceroxyleae) endemic to north Queensland and with its closest ally in the Andes: (a) habit; (b) inflorescences; (c) young infructescences; (d) habit in forest.

Fig. 6.8. *Arenga hastata* (Arecoideae, Caryoteae) a tiny, simply pinnate palm, endemic to lowland rain forest in West Malaysia. Here in Terengganu. Contrast Fig. 6.9.

Fig. 6.9. *Caryota maxima* (Arecoideae, Caryoteae) the Fish Tail Palm, a giant 40 m tall, doubly pinnate palm of lower montane rain forest throughout Sundaland. Contrast Fig. 6.8. (Genting Highlands, Malaysia.)

flowering. They are clearly very closely related. All genera are present in south-east Asia; *Wallichia* peters out in peninsular Thailand and beyond Wallace's line *Arenga* and *Caryota* are represented by only two or three species each. Evidence from present-day distribution suggests that these two genera have spread across Wallace's line in relatively recent time. Pollen referred to *Arenga* has been reported from the Lower Miocene of Borneo (Muller 1981) and this seems to support a west to east dispersal direction.

Tribe Areceae (5), like Tribe Calameae in subfamily Calamoideae, displays strong bicentricity. Unfortunately we still know little of the complex interrelationships between the subtribes of the Areceae. All the 84 genera have specialized pseudomonomerous gynoecia, except five which are tricarpellate. One of the exceptional tricarpellate genera, *Orania*, is present in Malesia. The genus occurs in Madagascar (1 species) (Dransfield and Uhl 1984), Sundaland (1 species, Fig. 6.10), Philippines (2 species), Sulawesi and Maluku (1 species) and New Guinea (9 species) (Essig 1980). (As mentioned above, the one Australian species has recently been shown not to belong to this genus.) Greatest diversity occurs in New Guinea. *Orania* is a relatively unspecialized genus and its distribution suggests a long history, but there is no fossil record and at present it is not possible to suggest whether the west Malesian taxa reached Sundaland via India or from across Wallace's line.

The pseudomonomerous members of the Arecoideae are divided among 11 subtribes of which

Fig. 6.10. *Orania sylvicola* (Arecoideae, Areceae, Oraniinae). *Orania* is the only Malesian tricarpellate genus of the subfamily, and is centred in New Guinea where there are nine species. (Ulu Gombak, Malaysia.)

Fig. 6.11. *Cyrtostachys renda* (Arecoideae, Areceae, Cyrtostachydinae), the Sealing Wax Palm, named from its crimson crownshafts. Inhabitant of the Sundaland peat-swamp forest. All the other 7 *Cyrtostachys* spp. come from Papuasia. (Singapore Botanic Garden.)

seven are present in Malesia or Australasia. However there is a strong Sahul bias; all seven subtribes are richly represented in the Sahul region and except for subtribe Arecinae the Sunda region has only a few representatives.

(a) Subtribe Cyrtostachydinae contains the single genus *Cyrtostachys*, with seven species in Papuasia and one in west Malesia (Fig. 6.11). A fossil pollen record of the genus from the Upper Miocene in Borneo (Muller 1968), if correct, would not contradict a theory of spread of the genus from the Sahul shelf.

(b) Subtribe Ptychospermatinae with eight genera, is entirely Papuasian or west Pacific in distribution except for one species of *Veitchia* in Palawan, Philippines; *Veitchia* itself is otherwise known only in Fiji and Vanuatu.

(c) Subtribe Iguanurinae, consists of 27 genera richly represented in New Caledonia, the west Pacific and New Guinea. There are a few exceptional genera. *Bentinckia* (2 species) occurs only in the Nicobar Islands and southern India, *Dictyosperma* (1 species) occurs in the Mascarene Islands, *Iguanura* (*c.*20 species, Fig. 6.12) is present only in Sundaland, *Heterospathe* (32 species) occurs mainly in New Guinea and the west Pacific with about 5 species in the Philip-

Fig. 6.12. *Iguanura polymorpha* (Arecoideae Areceae, Iguanurinae), small undergrowth palm of Sundaland rain forests. (Pahang, Malaysia.)

Fig. 6.13. *Oncosperma horridum* (Arecoideae Areceae, Oncospermatinae), from inland rain forests of Sundaland. (Ulu Gombak, Malaysia.)

pines, while *Rhopaloblaste* (7 species) occurs in New Guinea, the Solomons, and Moluccas with one species in Malaya and one in the Nicobar Islands. The extant distribution of the subtribe suggests a long Gondwanic history, with, perhaps, the rafting of a few taxa northwards with India into the northern hemisphere.

(d) Subtribe Oncospermatinae, very closely related to Iguanurinae, is, in contrast, distributed in the Seychelles, Mascarenes, Sri Lanka and Sundaland with one species of the widespread genus *Oncosperma* (Fig. 6.13) transgressing Wallace's line into Sulawesi. Rafting with India is also suggested.

(e) Subtribe Arecinae, in marked contrast, is almost equally divided between Sundaland and Papuasia. There is no transgression of the four Papuasian genera across Wallace's line while of the four western genera, only a few species of the large genus *Pinanga* (over 100 species, Fig. 6.14) are found east of Wallace's line, together with several species of the genus *Areca*. Despite this, the genera do seem to form a natural unit. Fossil pollen has been referred to *Nenga*, a Sundaic member of this subtribe, from the Lower Miocene of Borneo (Muller 1979). If this identity is correct then a western presence of the subtribe before the final juxtaposition of the two plates seems proved. On the other hand, the identity of a fossil pollen grain from the Maastrichtian of Cameroon with *Areca* (Muller 1981) seems highly questionable in view of the remarkable parallels in pollen structure within the palms (Ferguson, in press).

In summary, Tribe Areceae show a very strong austral bias, except for a few taxa which seem to have arrived in Sundaland prior to the formation of Malesia, possibly via rafting on the Indian plate, or by island-hopping across the area destined to become Malesia.

Tribe Cocoeae (6), the last to be considered, are entirely New World in distribution except for one genus in southern Africa, one in Madagascar, one recently extinct on Easter Island (Dransfield *et al.* 1984) and the coconut (Fig. 6.5). Gruezo and Harries (1984) and Harries (1984) have recently reported quite convincingly the presence of apparently wild coconuts in the Philippines and Queensland. This does not conflict with predictions (Harries 1979) based on the distribution of cultivars of this widespread crop. Fossils related to *Cocos* in New Zealand from the Pleiocene (Berry 1926) seem further to support the theory that the coconut may be of western Pacific origin.

Fig. 6.14. *Pinanga paradoxa* (Arecoideae, Areceae, Arecinae), tiny member of a very big largely Sundaic genus; all *Pinanga* spp. are palms of the rain forest undergrowth. (Ulu Sat, Taman Negara, Malaysia.)

THE ORIGIN OF THE MALESIAN PALM FLORA

From the above detailed analysis it seems possible to postulate a mixed origin of the Malesian palm flora from distinctly Laurasian and Tethyan elements from the north and west and distinctly Gondwanic elements from the south and east. Moore (1973*b*) proposed west Gondwanaland as the origin of the family. In contrast, we believe that the extant distribution of the least specialized apocarpic palms predominantly in the northern hemisphere, and the abundance of apparently Coryphoid macrofossils in the Mississippi embayment area (Daghlian 1981) at the very beginning of the fossil record of the palms, suggest that the family were as well developed in the Laurasian region as in Gondwanaland. We propose an origin of the family prior to, or at, the separation of Gondwanaland from Laurasia during the Cretaceous, with independent development in the northern and southern hemispheres. Although the earliest identifiable palm fossils are from the Upper Cretaceous, we suggest that the palms probably had an earlier origin. Some of the putative early monocotyledonous pollen recorded from the Potomac beds of the early Cretaceous of North America (Walker and Walker 1984) appear remarkably similar to some extant palm pollen types, but of course these rather unspecialized monosulcate grains are of widespread occurrence among monocotyledons. Although I do not wish to equate the fossils with palms, I do not believe there is sufficient evidence to discount their identity with palms.

By the late Oligocene the palm floras of the southern and northern hemispheres were probably already very diverse and many extant genera appear already to have evolved. The Sundaland palm flora appears at this time to have consisted of Laurasian palm elements, possibly enriched with Gondwanic elements which had rafted on India into Laurasia in the middle Eocene, or to have crossed over on Gondwanic fragments in the Malesian region (this volume, Chapter 2). The Sahul palm flora probably consisted entirely of Gondwanic elements. By the late Miocene the two palm floras were brought into close contact and since then they have intermingled to various extents. The large genus *Calamus* seems to have been particularly successful in colonizing the Sahul plate. There are obviously many problems yet to be elucidated, such as the origins of *Metroxylon* and *Pigafetta* and their relationships with the American Calamoid genera. It is hoped that when our phylogenetic analysis is completed we shall be in a better position to assess relationships and biogeography. However the palms do seem to show quite remarkable bicentricity in the Malesian region, to a greater degree than almost any other angiosperm family.

ACKNOWLEDGEMENTS

The photographs illustrating this chapter are from the collection of T. C. Whitmore except Fig. 6.7 which is from *Principes* 29(2) 1985, 58.

7 THE MAMMALS OF SULAWESI

Guy G. Musser

Sulawesi and nearby offshore islands have 122 native species of mammals (55 genera, 15 families). All are placentals of Sundaic origin except for 3 species of the marsupial genus Phalanger. *The non-volant mammals are shown to be of ancient origin, many are most primitive of their lineage. Origin is believed to have been across sea, for the fauna is the unbalanced, depauperate one of an oceanic island (40 per cent of species are bats), with many Sundaic families absent. Fossil vertebrates are all from the south-west arm and of either Pliocene or Pleistocene age. It is speculated that arm was an island whose fauna became exterminated by competition when it became joined to central Sulawesi.*

THE NATIVE MAMMALS

Phalangers

Three species of phalangers, *P. ursinus*, *P. pelengensis*, and *P. celebensis*, comprise 2 per cent of the total indigenous mammal fauna in Sulawesi and nearby islands.

The bear phalanger, *Phalanger ursinus*, is the largest in body size. It has been recorded from lowlands and mountain forests on mainland Sulawesi, the south-eastern islands of Muna and Butung, Pulau Peleng, Pulau Melenge in the Togian islands and Pulau Lirung in the Talaud islands to the north and east of north-eastern Sulawesi. *Phalanger celebensis* (Fig. 7.1) occurs on mainland Sulawesi and the Sangihe islands. Specimens are from lowlands and middle altitudes within the ranges of tropical lowland and evergreen lower montane rain forests. *Phalanger pelengensis* is known only from the islands of Peleng and Taliabu.

Subfossil specimens of both *P. ursinus* and *P. celebensis* have been identified from cave deposits in the south-western peninsula of Sulawesi (Hooijer 1950; Clason 1976); none of the material comes from deposits older than about 8000 years BP and most comes from younger strata.

Shrews

Native living *Soricidae* on Sulawesi are represented by one genus, *Crocidura* (Fig. 7.2), and eight species, or six per cent of the total native mammal fauna. The information listed in Table 7.1 provides a general picture of their geographical and ecological distributions.

Subfossils of Sulawesian *Crocidura* are meagre. I have identified a partial left dentary from Batu Edjaja as *C. rhoditis*. The specimen was obtained during the archaeological expedition to the Makassar region reported by Mulvaney and Soejono (1970).

Bats

Fifty-nine species of bats in six families are recorded from Sulawesi, nearly half of the native mammals known from the area (Table 7.2). Three genera and nine species are endemic, which accounts for 13 per cent of the endemic mammal fauna. The remainder of the species found on mainland Sulawesi and nearby islands also occur in different combinations elsewhere within the Indo–Australian region.

The known fossil record does not help to expand our knowledge of the Sulawesian bat fauna. Unidentified subfossil fragments of mandibles and isolated teeth are reported from Ulu Leang I in the Makassar region (Clason 1976). Dentary fragments obtained by Mulvaney and Soejono (1970) from Leang Burung in south-western Sulawesi have been identified as species that still live on the island.

Fig. 7.1. *Phalanger celebensis*, one of the three species of *Phalanger* native to Sulawesi. (Colour plate 15 in Meyer 1896).

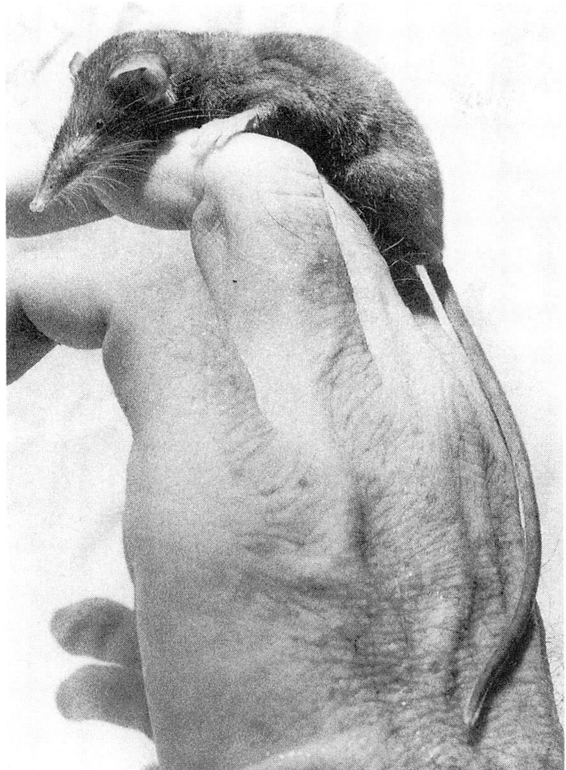

Fig. 7.2. An adult female of the long-tailed shrew *Crocidura elongata*, caught in lowland forest. In contrast to *C. nigripes*, the long-tailed shrew occurs at all altitudes from lowlands to mountain tops.

Tarsiers

Two species of tarsiers, which makes up 2 per cent of the total native mammalian fauna, live in the forests of Sulawesi.

Tarsius spectrum is the common tarsier on the mainland and nearby islands (Niemitz 1984*a*, 1985; Musser and Dagosto, ms.). Specimens of *T. spectrum* are from most major regions of Sulawesi, including the south-eastern peninsula (which is not indicated on the map in Niemitz 1984*a*). All were collected in tropical lowland evergreen rain forest at low and middle altitudes. *Tarsius pumilus* lives in the cool and wet moss forests of central Sulawesi. The species is much smaller in body size than *T. spectrum*, and is also distinguished from it by conspicuous dental and external features.

Macaques

Aside from tarsiers, the only other native non-human primates are species of *Macaca*, which comprise 3 per cent of the total species of mammals native to Sulawesi and 6 per cent of the total endemic species. Our modern picture of the morphological and phylogenetic relationships among their populations, as well as their possible evolutionary histories, comes primarily from the works of Fooden (1969), Albrecht (1978), and Groves (1980*b*).

Four species are currently recognized. *Macaca nigra* (*nigrescens* as a subspecies) occurs in the

Table 7.1
Distributions of native *Crocidura* on Sulawesi (based upon my study of specimens)

Species	*North*	*Central*	*South-east*	*South-west*	*Tropical rain forest*	*Activity time*
C. elongata	+	+			Lowland evergreen to moss[3]	Nocturnal
C. nigripes[1]	+	+			Lowland evergreen	Nocturnal
C. rhoditis	+	+		+	Lowland evergreen to moss	Nocturnal
C. lea	+	+			Lowland evergreen to moss	Nocturnal
C. levicula		+	+		Lowland evergreen to moss	Diurnal
C. sp. *A*		+			Lowland evergreen to moss	Diurnal
C. sp. *B*		+			Moss	Nocturnal
C. sp. *C*[2]					Lowland evergreen	?

[1] Also recorded from Pulau Lembeh, north-eastern Sulawesi.
[2] From Pulau Peleng and Pulau Taliabu (Sula Islands).
[3] I use moss forest to apply to upper and lower montane rain forest (see Whitmore 1975).

Table 7.2

Geographical distributions of bats native to Sulawesi and nearby islands. Records are from Koopman (1979, 1982, 1984, and unpublished ms.), Musser *et al.* (unpublished ms.), Hill (1983), Rookmaaker and Bergmans (1981), Bergmans (1978, 1979), de Jong and Bergmans (1981), Bergmans and Rozendaal (1982), Rozendaal (1984), Laurie and Hill (1954), and unpublished museum collections

Taxon	Sulawesi endemic	Malaya[1]	Sumatra[2]	Java, Bali[3]	Borneo[4]	West Sumatran islands[5]	Lesser Sunda islands[6]	Philippines	Sula islands	Taulad islands	Moluccas	New Guinea region[7]	Australia[8]
Pteropodidae													
Rousettus amplexicaudatus		+	+	+	+	+	+	+			+	+	
Rousettus celebensis	+												
Boneia bidens	+												
Pteropus hypomelanus		+			+	+		+		+	+	+	
Pteropus griseus							+						
Pteropus caniceps									+		+		
Pteropus personatus											+		
Pteropus alecto				+			+						+
Acerodon celebensis									+				
Neopteryx frosti	+												
Styloctenium wallacei	+												
Dobsonia crenulata											+		
Dobsonia exoleta	+												
Harpyonycteris whiteheadi								+					
Cynopterus brachyotis		+	+	+	+	+		+		+			
Chironax melanocephalus		+	+	+									
Thoopterus nigrescens											+		
Nyctimene minutus											+		
Nyctimene cephalotes							+				+	+	
Eonycteris spelaea		+	+	+	+		+	+					
Macroglossus minimus		+	+	+	+		+	+			+	+	+
Emballonuridae													
Emballonura alecto					+			+					
Emballonura raffrayana											+	+	
Emballonura nigrescens											+	+	
Taphozous melanopogon		+	+	+	+		+			+			
Saccolaimus saccolaimus		+	+	+	+		+			+	+	+	
Megadermatidae													
Megaderma spasma		+	+	+	+	+		+	+				
Rhinolophidae													
Rhinolophus celebensis				+			+						
Rhinolophus philippinensis					+		+	+					+
Rhinolophus euryotis											+	+	
Hipposideros bicolor		+	+	+		+	+	+					
Hipposideros ater			+	+	+			+				+	
Hipposideros cervinus		+	+		+	+		+			+	+	+
Hipposideros diadema		+	+	+	+	+	+	+			+	+	+
Hipposideros dinops												+	
Hipposideros inexpectatus	+												
Vespertilionidae													
Myotis muricola		+	+	+	+		+	+			+		
Myotis adversus		+	+	+	+								+
Myotis horsefieldi		+		+	+			+					
Myotis formosus			+	+				+					

Table 7.2 (*cont.*)

Taxon	Sulawesi endemic	Malaya[1]	Sumatra[2]	Java, Bali[3]	Borneo[4]	West Sumatran islands[5]	Lesser Sunda islands[6]	Philippines	Sula islands	Taulad islands	Moluccas	New Guinea region[7]	Australia[8]
Pipistrellus javanicus		+	+	+	+			+					
Pipistrellus tenuis		+	+	+	+		+	+			+	+	+
Pipistrellus petersi								+			+		
Pipistrellus minahassae	+												
Tylonycteris robustula		+	+	+	+		+						
Hesperopterus gaskelli	+												
Scotophilus kuhli		+		+	+		+	+					
Scotophilus celebensis	+												
Miniopterus australis					+						+	+	+
Miniopterus pusillus			+	+	+		+	+			+	+	
Miniopterus fuscus		+		+				+					
Miniopterus schreibersi			+	+			+				+	+	+
Miniopterus tristis								+				+	
Murina florium							+				+	+	+
Kerivoula hardwickei		+	+	+	+	+	+						
Kerivoula papillosa		+	+	+	+								
Kerivouls jagorii				+	+			+					
Molossidae													
Mops sarasinorum								+					
Cheiromeles parvidens								+					
Totals	9	22	22	27	25	8	20	26	3	2	23	17	2

[1] Includes surrounding islands on the Sunda shelf east to the Anambas and North Natunas.
[2] Includes adjacent islands on the Sunda shelf.
[3] Includes nearby islands on the Sunda shelf.
[4] Includes adjacent islands on the Sunda shelf, as well as the Palawan-Calamianes group of the Philippines.
[5] The Mentawais and other west Sumatran islands off the Sunda shelf, including Pulau Nias.
[6] Includes Nusa Penida.
[7] Includes the Bismarcks, Vanuatu, and Solomons.
[8] Includes Tasmania.

north-eastern arm west to the Gorontolo area; *M. tonkeana* (*hecki* as a subspecies) occupies the rest of northern Sulawesi and all of the central part of the island; *M. ochreata* (*brunnescens* as a subspecies) is found on mainland south-eastern Sulawesi as well as the large islands of Muna and Butung; and *M. maura* is the macaque of the south-western peninsula.

Sulawesian macaques are forest dwellers mainly of the lowlands. We often saw troops of half a dozen individuals and now and again would encounter troops of 15–20 monkeys feeding and sleeping in tall strangler figs.

Squirrels

Three genera—*Rubrisciurus*, *Hyosciurus*, and *Prosciurillus*—and at least six species of squirrels are endemic to Sulawesi. The species comprise 5 per cent of the total mammalian fauna and 9 per cent of the endemic component. Five are characterized by conspicuous and discrete morphological features.

Rubrisciurus rubriventer is the largest squirrel on the island and probably occurs wherever suitable lowland forest is present. It is adapted to exploit resources on the ground and in the low understory.

Resources on the floor of the forest are also utilized by two species of *Hyosciurus*, the extraordinary Sulawesian long-nosed ground squirrel (Laurie and Hill 1954). *Hyosciurus heinrichi* is common in moss forest from mountain summits down to the bottom part of lower montane forest at 1500 m. It is known only from the mountains of central Sulawesi. *Hyosciurus ileile*, which is larger in body size and has a shorter snout, replaces the other species in forest from 1500 m down to the coast, and is uncommon.

Prosciurillus is the third endemic squirrel genus and contains species of tree squirrels in three body sizes: small, medium, and intermediate. *P. murinus*, the smallest sciurid native to the island, occurs in central Sulawesi, the northern peninsula (including Pulau Lembeh), the south-eastern arm, and the south-western peninsula.

Of all the native sciurids, less is known about *P. abstrusus* (Moore 1958) than any of the others. It was collected at 1500 m in montane rain forest on the Mengkoka mountains in the south-eastern peninsula of Sulawesi. There it occurs together with *P. murinus* and *P. leucomus*.

Prosciurillus leucomus is of medium body size and has a long bushy tail. It is the canopy squirrel in the Sulawesi region. This squirrel lives in primary forest at all altitudes from sea level to mountain tops.

Nowhere have all six species of sciurids been collected together in the same forest, but they do co-occur in different combinations. In central Sulawesi, for example, we encountered *Hyosciurus heinrichi*, *Prosciurillus murinus*, and *P. leucomus* in moss forest. At altitudes in lowland forest, *Rubrisciurus rubriventer* and *Hyosciurus ileile* were found in the same forest as *Prosciurillus murinus* and *P. leucomus*.

Rats and mice

In their checklist of the land mammals of New Guinea, Celebes and adjacent islands, Laurie and Hill (1954) recognized six genera and 35 species in the family Muridae from Sulawesi and nearby islands. Twenty-nine of the species were listed in the genus *Rattus*. My latest count is 36 species distributed among 14 genera with only four in *Rattus* (Table 7.3).

All the native murids are endemic to Sulawesi and make up 30 per cent of the total mammalian fauna and 52 per cent of all the endemic species. Of the 14 genera, four are not confined to Sulawesi. *Crunomys* is also represented in the Philippine islands (Musser 1982), *Hearomys* includes species on Palawan and Borneo, *Maxomys* embraces species from Sundaland and Indochina, and *Rattus* contains species distributed throughout the Indo–Australian region (Musser and Newcomb 1983).

During the last 30 years information has been gathered about the modern rats and mice native to Sulawesi but nothing is known of the species living during periods older than 8000 to 9000 years BP. We know only that nine of the species occurring in the south-western peninsula are represented by subfossil samples—younger than 9000 BP—from cave deposits in that region. Seven of the nine species are also represented by modern samples from the south-western arm, and two, *Taeromys celebensis* and *T. punicans*, have not yet been collected there but are found elsewhere in Sulawesi (Table 7.3; Musser 1984).

Civets

Three viverrids have been recorded from Sulawesi (Laurie and Hill 1954) but only *Macrogalidia musschenbroekii*, the Sulawesi palm civet, is native. It is also endemic and the only member of the order Carnivora native to the island. The Malay civet, *Viverra tangalunga*, and the common palm civet, *Paradoxurus hermaphroditus*, are not members of the indigenous fauna (Table 7.4).

Macrogalidia has also been identified in samples from cave deposits at Toalian sites in south-western Sulawesi (Hooijer 1950).

Pigs

The Sulawesi wild pig, *Sus celebensis*, and the babirusa (pig deer), *Babyrousa babyrussa*, are the two endemic species of Suidae now living on Sulawesi. Together they form 2 per cent of the

Table 7.3

Distributions of native rats and mice on Sulawesi. Most of the species listed are defined and described in reports listed in Musser (1984) and Musser and Newcomb (1983); others, including undescribed taxa, will be discussed in manuscripts being prepared for publication

Taxon	*North*	*Central*	*South-east*	*South-west*	*Tropical rain forest*
Crunomys celebensis		+			Lowland evergreen
Echiothrix leucura	+	+			Lowland evergreen
Tateomys macrocercus		+			Moss[1]
Tateomys rhinogradoides		+			Moss
Melasmothrix naso		+			Moss
New Genus and Species		+			Moss
Margaretamys beccarii	+	+			Lowland evergreen
Margaretamys elegans		+			Moss
Margaretamys parvus		+			Moss
Lenomys meyeri	+	+		+[2]	Lowland evergreen to moss
Eropeplus canus		+			Moss
Haeromys sp.		+			Moss
Haeromys minahassae	+	+			Lowland evergreen
Maxomys muschenbroekii	+	+	+	+[2]	Lowland evergreen to moss
Maxomys dollmani		+	+		Moss
Maxomys hellwaldii	+	+	+	+[2]	Lowland evergreen
Maxomys sp. *A*		+			Lowland evergreen
Maxomys sp. *B*		+			Lowland evergreen to moss
Maxomys sp. *C*		+			Moss
Bunomys chrysocomus	+	+	+	+	Lowland evergreen to moss
Bunomys fratrorum	+				Lowland evergreen to moss
Bunomys andrewsi		+	+	+[2]	Lowland evergreen
Bunomys penitus		+	+		Moss
Bunomys sp. *A*		+			Lowland evergreen
Bunomys sp. *B*		+			Moss
Paruromys dominator	+	+	+	+[2]	Lowland evergreen to moss
Taeromys celebensis	+	+	+	+[3]	Lowland evergreen
Taeromys taerae	+				Lowland evergreen
Taeromys arcuatus		+	+		Moss
Taeromys hamatus		+			Moss
Taeromys punicans		+		+[3]	Lowland evergreen
Taeromys callitrichus	+	+	+		Lowland evergreen to moss
Rattus xanthurus	+	+	+	+[2]	Lowland evergreen to moss
Rattus hoffmanni	+	+	+	+[2]	Lowland evergreen to moss
Rattus marmosurus	+				Moss
Rattus sp.		+			Lowland evergreen
Total	15	33	12	10	
Per cent of total spp.	42	92	33	28	

[1] I use moss forest to apply to upper and lower montane rain forest (see Whitmore 1975).
[2] Known by samples from extant populations and subfossils.
[3] Known only by samples of subfossils.

total mammalian fauna and 3 per cent of the endemics. Specimens of the Sulawesi wild pig come from all major peninsulas, the central region, and the islands of Lembeh, Peleng, Butung, and Selayar (see the distribution map in Groves 1981). The species also occurs on Halmahera in the Moluccas; Flores, western Timor, Pulau Partiti, and Pulau Landu, in Nusa Tenggara; and Pulau Simaleue off the west coast of northern Sumatra. Grove (1981) considers these records to represent feral populations derived from introductions.

The species is known from the south-western peninsula only by subfossil fragments (Hooijer 1950, Groves 1980*a*; the isolated teeth originally described as *B. babyrussa beruensis* from Pleisto-

Table 7.4
Mammals introduced to Sulawesi

Taxon	*Notes*
Phalangeridae	
Phalanger maculatus	Known only from Pulau Selayar and indistinguishable from *P. maculatus chrysorrhos* of Moluccas and New Guinea
Soricidae	
Suncus murinus	Widespread in Asia; on Sulawesi occurs in villages, towns and cities
Sciuridae	
Callosciurus prevostii	Known from only two spots in northern Sulawesi (Laurie and Hill 1954)
Callosciurus notatus	Recorded only from Pulau Selayer and morphologically indistinguishable from Javan *C. notatus*
Muridae	
Rattus argentiventer	Rice fields and scrub at low and middle altitudes
Rattus nitidus	Known only from mountain valleys of central Sulawesi
Rattus norvegicus	Restricted to large port cities
Rattus exulans	Common everywhere in human habitats and disturbed forest near villages
Rattus rattus rattus	Uncommon in Makassar region; not established
Rattus rattus (*palelae*, *dammermani*, and *pelengensis*)	Lowlands to middle altitudes in scrub, gardens, fields, villages and towns on mainland and nearby islands
Mus musculus castaneus	Buildings in towns and cities
Hystricidae	
Hystrix javanica	Recorded only from south-western peninsula
Viverridae	
Paradoxurus hermaphroditus	Very few records from north-eastern and south-western peninsulas
Viverra tangalunga	Common around villages and in parts of primary forests
Cervidae	
Cervus timorensis	Widespread in suitable habitat; native to Java

cene deposits, Hooijer 1948, are examples of *Celebochoerus*—see Hooijer 1954, 1958). Babirusa also occurs on the islands of Taliabu and Sulabesi in the Sula islands and on Buru; Groves (1980*a*, 1981) believes these populations to represent human introductions during historic times.

South-western Sulawesi has also yielded subfossils from cave deposits and part of a skull reputed to be Pleistocene in age of *Sus celebensis* (Hooijer 1950, 1969). Morphology and size of the elements are very close to those in samples of living pigs.

Anoas

The latest and best taxonomic study of Sulawesi dwarf buffalo is that by Groves (1969) who studied most of the museum specimens. He interpreted the morphological variation found within these samples to represent two species. The larger one with triangular horns is *Bubalus depressicornis*, the smaller anoa with conical horns is *B. quarlesi*; they appear to be mainly lowland and montane in habitat respectively. Both are endemic to Sulawesi and part of a special fauna adapted to life in primary forests.

Bubalus depressicornis has been recorded from the northern, central, and south-eastern parts of Sulawesi, and although there are no modern records from the south-western peninsula, the species is represented by subfossils associated with the Toalian culture (Hooijer 1950), and by reputedly Pleistocene fragments (Hooijer 1948, 1972; Sartono 1979).

Bubalus quarlesi is the only anoa we saw in central Sulawesi. It is also known by specimens taken in the northern peninsula, the south-eastern peninsula and Pulau Butung, and the south-western arm. Animals or signs of them were sighted at all altitudes from *c.* 100 m up to the summit of Gunung Nokilalaki. Samples of *B. quarlesi* are also part of the cave deposits from the Toalian sites (Hooijer 1950).

THE INTRODUCED MAMMALS

Throughout the Indo–Australian region, the mammalian fauna of any one place consists of truly native species and those which have been intentionally or inadvertently introduced by humans. Separating out the non-natives is a crucial step to understanding the distributional and phylogenetic patterns formed by the native species. The fauna of Sulawesi is no exception. In addition to the 122 native species, there are 15 that are not part of the indigenous fauna (Table 7.4).

Most of the introduced species live in habitats associated with human modification of primary forests and all are non-volant; they all occur also in regions well outside of Sulawesi. In striking contrast, all the non-volant species native to Sulawesi are tied to forest and nearly all are endemic.

In addition to the introduced mammals listed in Table 7.4, domesticated water buffalo, Bali cattle, pigs, cats, dogs, and goats all have the potential of becoming feral, and feral populations of dogs, cats, and buffalo can now be found on Sulawesi (personal observations).

DISTRIBUTIONAL PATTERNS OF ENDEMICS

The mammals show two sets of distributional patterns: species restricted to particular geographical regions, and ranges of species associated with altitude.

Regional distribution

Certain species are widespread on the island wherever suitable habitat exists. Examples are the two phalangers; probably *Crocidura nigripes* and *C. rhoditis*; some bats, such as *Rousettus celebensis* and *Cynopterus brachyotis*; *Tarsius spectrum*; the squirrels, *Rubrisciurus rubriventer* and *Prosciurillus murinus*; many species of rats and mice (Table 7.3); *Sus celebensis* and babirusa; the two kinds of anoa; and probably the native civet.

Some species appear to be restricted to one of the peninsulas, central Sulawesi, or some off-

Table 7.5
Regional endemic distributions of native Sulawesian mammals

North-east	*Central*	*South-east*	*South-west*	*Insular*
Bonia bidens	*Crocidura* sp. *B*	*Macaca ochreata*	*Macaca maura*	*Phalanger pelengensis* (Peleng)
Macaca niger	*Neopteryx frosti*	*Prosciurillus abtrusus*		*Crocidura* sp. *C* (Peleng, Taliabu)
Bunomys fratrorum	*Tarsius pumilus*			*Dobsonia crenulata* (Sangihe, Togian)
Taeromys taerae	*Macaca tonkeana*			*Rattus* species (Peleng)
Rattus marmosurus	*Hyosciurus heinrichi*			
	Bunomys sp. *A*			
	Bunomys sp. *B*			
	Taeromys hamatus			
	Crunomys celebensis			
	Tateomys macrocercus			
	Tateomys rhinogradoides			
	Melasmothrix naso			
	Genus and species nov.			
	Margaretamys elegans			
	Margaretamys parvus			
	Eropeplus canus			
	Haeromys species			
	Maxomys sp. *A*			
	Maxomys sp. *B*			
	Maxomys sp. *C*			

Fig. 7.3. *Echiothrix leucura*, the large-bodied shrew rat of tropical lowland evergreen rain forest on Sulawesi (from colour plate 9 in Meyer 1898).

shore islands. This pattern reflects regional endemism (Table 7.5). Eventually we may be able to include more species of shrews and bats in endemic clusters, and exclude some rats from the list for central Sulawesi. It is unlikely for example, that some rats, such as species of *Tateomys* and *Melasmothrix*, will be found in the south-eastern arm because the mountains there are not high enough to support moss forest habitats. On the other hand, a species like *Crunomys celebensis* is known by so few specimens and is so difficult to collect that it may live elsewhere but has been overlooked.

Regional endemism exists also at the level of subspecies. This pattern is most evident in rats and involves the north-eastern and south-western arms. Samples of *Echiothrix leucura* (Fig. 7.3) from the north-east, for example, are morphologically distinct from samples taken in central Sulawesi. Specimens of *Maxomys hellwaldii*, *Bunomys chrysocomus*, *Bunomys andrewsi*, *Paruromys dominator*, *Rattus xanthurus*, and *Rattus hoffmanni* from the south-western peninsula can be separated by morphological features from samples obtained in other regions (Musser 1984). The babirusa may be an example of a large-bodied mammal whose populations show regional differentiation on the island. Groves (1980*a*) indicated that the small samples from central Sulawesi available to him contained much smaller pigs compared with samples from the north-eastern arm.

Insular subspecific endemism is also present in some mammals. *Phalanger celebensis*, *Tarsius spectrum*, and *Prosciurillus leucomus* from the Sangihe islands are all morphologically distinct from mainland populations. In external and cranial features, *Phalanger ursinus* and *Rattus xanthurus* on Pulau Peleng contrast significantly with mainland populations.

FAUNAL RELATIONSHIPS

The Sulawesian fauna is unlike that of the Philippines (Fooden 1969, Groves 1976, Heaney ms.) and I find no close relationship between the faunas of Sulawesi and the Lesser Sunda Islands. Other than recently introduced mammals, the only species common to the three groups of islands are bats. Three species are shared by the Philippines and Sulawesi and only one by the Lesser Sundas and Sulawesi (Table 7.2). A few genera are shared by the three areas. *Rattus*, for example, is native on the Philippines and Sulawesi, but the species are not closely related and are tied together only because they probably originated independently from ancestral populations on the Sunda shelf. Of the non-*Rattus* genera native to the backbone of the Philippine islands, only *Crunomys* is shared by the Philippines and Sulawesi (Musser *et al.* 1985) but the Sulawesian species is morphologically very different from the Philippine forms (Musser 1982) and also may reflect independent origin from the west. Phylogenetic relationships among the rest of the native Philippine murids are still being worked out but they do not tie to the Sulawesian fauna (personal observations). The rest of the Philippine native mammals, especially those in the Mindanao Faunal Province, can be derived from the Sunda shelf and probably from Borneo (Heaney ms.). Groves (1976) pointed out that the Sangihe and Talaud islands, which lie between Sulawesi and Mindanao, were devoid of mammals with affinities to the Philippines; the known species were either widespread (bats and commensals) or directly related to Sulawesi or the Moluccas. No new evidence challenges this view.

The mammals of the Lesser Sunda Islands are still poorly known. Many of the endemic species consist of murids and are known only by subfossil fragments. This does not mean that most of the fauna is extinct, only that islands, especially Flores and Komodo, have been inadequately explored to determine the kinds of native rats and mice still living in existing habitats. Still, of the known fauna, none is directly related to natives living on Sulawesi (Musser 1981).

Where does the Sulawesian fauna fit in the context of mammalian species distributions in that vast region from the Sunda to Sahul shelves where the western boundary is characterized by a fauna of Asian origin and the eastern boundary by one of basically Gondwanic derivation, with some elements bridging the two areas? Answers come from comparing the Sulawesi fauna with those on the Sunda shelf to the west and island groups and continent to the east.

The overall fauna

Actual numbers of native and endemic species found in eight major regions from the Sunda shelf to New Guinea and Australia are listed in Table 7.6. The number of species in a family expressed in per cent of the total fauna of an area, and the number of endemics relative to the total species in a family are listed in Table 7.7.

Several patterns are reflected by those data.

(1). The first is the distribution of living monotremes, which do not extend west of Australia and New Guinea, and the abrupt decrease in marsupials from Australia to Sulawesi, which is the western limit of this old mammalian group that is so characteristic of New Guinea and Australia.

(2). The second pattern complements the first. Species in 22 out of the 36 families of placentals found on the Sunda shelf do not range eastward beyond the continental margin. Representatives of these families in Sundaland are Asian and some have distant relatives in the Palaearctic region and Africa. The eastern decrease in Asian groups is also seen in six families that occur on the Sunda shelf and reach Sulawesi but no farther: tarsiers, macaques, squirrels, civets, pigs, and buffalo. The shrews are also a western group and a large cluster gets to Sulawesi. *Crocidura* sp. *C* occurs on the Sula islands and the Asian *C. maxi* is found in the Moluccas (Ambon). In this part of the world, Sulawesi is clearly the eastern point reached by most species of soricids.

(3). Seven families have native members ranging from the Sunda shelf to New Guinea and Australia, and these distributions form the third pattern. Six of the families are bats (Pteropodidae, Emballonuridae, Megadermatidae, Rhinolophidae, Vespertilionidae, and Molosside) and the seventh is native rats and mice (Muridae).

(4). A fourth pattern reflects the distributions of families in each of the geographical regions. No marsupials or monotremes occur on the Sunda shelf. Only 2 per cent of the Sulawesian fauna is composed of marsupials, nearly half are bats (48 per cent), almost a third consists of rats and mice, and the remaining 20 per cent of the fauna contains species in seven other families that are definitely Asian in origin. Phalangers, a shrew, bats, and rats are on the Sula islands, 33 per cent of the species are bats, and 33 per cent are rats. Sixty per cent of the Talaud islands fauna consists of bats, the rest are murids (20 per cent) and phalangers (20 per cent). In the Moluccas the marsupials increase in diversity and form 8 per cent of the fauna, the bats comprise 80 per cent, species of rats and one shrew the rest. New Guinea supports monotremes and marsupials (32 per cent), bats (37 per cent), and rats and mice (31 per cent). The Bismarcks and Solomons have a large bat element (77 per cent), which seems to be characteristic of island groups in the Indo–Australian region, and few rats and marsupials. More than half (53 per cent) of the Australian fauna consists of monotremes and marsupials; the rest are bats (only 24 per cent) and murid rodents (23 per cent).

(5). A fifth pattern is revealed by the native and endemic bat distributions. Only 15 per cent of the total chiropteran fauna is endemic to Sulawesi, a value similar to that for the Sunda shelf (16 per cent). Relative endemicity is greater in the Talaud islands (33 per cent), New Guinea (25 per cent), Bismarcks and Solomons (33 per cent), and Australia (38 per cent). By contrast, only 2 per cent (1 species) of the Moluccan bats are endemic, and no endemics have been recorded from the Sula islands.

(6). The last pattern points again to the close faunal relationship between the Sunda shelf and Sulawesi and the distant link between that island and the New Guinea–Australian area. This pattern also emphasizes Sulawesi's isolation in terms

Table 7.6
Occurrences of modern, native species in families from the Sunda shelf to Australia. Actual number of species is denoted for each family followed by number of endemics in parentheses

Family	*Sunda shelf*[1]	*Sulawesi*	*Sula islands*	*Talaud islands*	*Moluccas*	*New Guinea*	*Bismarck and Solomons*[2]	*Australia*[3]
MONOTREMES								
Ornithorhynchidae								1 (1)
Tachyglossidae						2 (1)		1 (0)
MARSUPIALS								
Dasyuridae						13 (11)		39 (37)
Myrmecobiidae								1 (1)
Thylacinidae						1 (0)		1 (0)
Peramelidae					1 (1)	9 (6)	1 (0)	9 (7)
Thylacomyidae								2 (2)
Notoryctidae								1 (1)
Phalangeridae		3 (1)	2 (0)	1 (0)	3 (1)	9 (6)	1 (0)	6 (4)
Petauridae					1 (0)	14 (12)	1 (0)	11 (9)
Burramyidae						2 (1)		6 (5)
Potoroidae								9 (9)
Macropodidae						12 (8)	1 (0)	39 (37)
Tarsipedidae								1 (1)
Vombatidae								3 (3)
Phascolarctidae								1 (1)
PLACENTALS								
Erinaceidae	2 (1)							
Talpidae	1 (0)							
Soricidae	9 (3)	7 (7)	1 (1)		1 (0)			
Tupaiidae	12 (11)							
Cynocephalidae	1 (0)							
Pteropodidae	19 (1)	21 (5)	2 (0)	3 (1)	24 (1)	27 (9)	23 (13)	8 (0)
Rhinopomatidae	1 (0)							
Emballonuridae	6 (0)	5 (0)			5 (0)	6 (0)	4 (0)	7 (2)
Nycteridae	2 (1)							
Megadermatidae	2 (0)	1 (0)	1 (0)		1 (0)			1 (1)
Rhinolophidae	40 (6)	9 (1)			9 (0)	15 (2)	10 (1)	8 (1)
Vespertilionidae	33 (7)	21 (3)			13 (0)	17 (5)	8 (1)	28 (17)
Molossidae	6 (2)	2 (0)			1 (0)	6 (2)	1 (0)	6 (1)
Lorisidae	1 (0)							
Tarsiidae	1 (1)	2 (2)						
Cercopithecidae	15 (11)	4 (4)						
Pongidae	4 (4)							
Manidae	1 (0)							
Leporidae	1 (1)							
Sciuridae	54 (36)	6 (6)						
Rhizomyidae	2 (0)							
Muridae	47 (40)	36 (36)	3 (3)	1 (0)	7 (5)	60 (53)	10 (8)	56 (49)
Hystricidae	5 (2)							
Canidae	1 (0)							
Ursidae	1 (0)							
Mustelidae	11 (2)							
Viverridae	15 (5)	1 (1)						
Herpestidae	4 (2)							
Felidae	9 (2)							
Elephantidae	1 (0)							
Tapiridae	1 (0)							
Rhinocerotidae	2 (0)							
Suidae	3 (0)	2 (2)						
Tragulidae	2 (0)							
Cervidae	5 (2)							
Bovidae	3 (0)	2 (2)						
Totals	322 (139)	122 (69)	9 (4)	5 (1)	66 (8)	193 (116)	60 (23)	245 (189)

[1] Includes all islands on the Sunda shelf north to the Isthmus of Kra, as well as Pulau Nias, Pulau Enggano, and the Mentawai islands.

Table 7.7

Distributions of modern, native species in families from the Sunda shelf to Australia. Per cent of total fauna is listed followed in parentheses by per cent of endemic species within each family. Boundaries of geographical regions are explained in Table 7.6

Family	*Sunda shelf*	*Sulawesi*	*Sula islands*	*Talaud islands*	*Moluccas*	*New Guinea*	*Bismarck and Solomons*	*Australia*
MONOTREMES								
Ornithorhynchidae								< 1 (100)
Tachyglossidae						1 (50)		< 1 (0)
MARSUPIALS								
Dasyuridae						7 (85)		15 (87)
Myrmecobiidae								< 1 (100)
Thylacinidae						1 (0)		1 (0)
Peramelidae					2 (100)	5 (67)	2 (0)	4 (78)
Thylacomyidae								1 (100)
Notoryctidae								< 1 (100)
Phalangeridae		2 (33)	22 (0)	20 (0)	2 (100)	5 (67)	2 (0)	2 (67)
Petauridae					2 (0)	7 (86)	2 (0)	4 (82)
Burramyidae						1 (50)		2 (83)
Potoroidae								4 (100)
Macropodidae						6 (67)	2 (0)	16 (95)
Tarsipedidae								< 1 (100)
Vombatidae								1 (100)
Phascolarctidae								< 1 (100)
PLACENTALS								
Erinaceidae	1 (50)							
Talpidae	< 1 (0)							
Soricidae	3 (33)	6 (100)	22 (100)		2 (0)			
Tupaiidae	4 (92)							
Cynocephalidae	< 1 (0)							
Pteropodidae	6 (5)	17 (24)	33 (0)	60 (33)	36 (4)	13 (33)	38 (57)	3 (0)
Rhinopomatidae	< 1 (0)							
Emballonuridae	2 (0)	4 (0)			8 (0)	3 (0)	3 (0)	3 (29)
Nycteridae	1 (50)							
Megadermatidae	1 (0)	1 (0)	11 (0)		2 (0)			< 1 (100)
Rhinolophidae	12 (15)	7 (11)			14 (0)	8 (40)	17 (10)	3 (13)
Vespertilionidae	10 (21)	17 (14)			20 (0)	9 (29)	13 (13)	11 (61)
Molossidae	2 (33)	2 (0)			2 (0)	3 (33)	2 (0)	2 (17)
Lorisidae	< 1 (0)							
Tarsiidae	< 1 (100)	2 (100)						
Cercopithecidae	5 (73)	1 (100)						
Pongidae	1 (100)							
Manidae	< 1 (0)							
Leporidae	< 1 (100)							
Sciuridae	17 (65)	5 (100)						
Rhizomyidae	1 (0)							
Muridae	15 (85)	30 (100)	33 (100)	20 (0)	11 (71)	31 (88)	17 (80)	23 (88)
Hystricidae	2 (40)							
Canidae	< 1 (0)							
Ursidae	< 1 (0)							
Mustelidae	3 (18)							
Viverridae	5 (33)	1 (100)						
Herpestidae	1 (50)							
Felidae	3 (22)							
Elephantidae	< 1 (0)							
Tapiridae	< 1 (0)							
Rhinocerotidae	1 (0)							
Suidae	1 (0)	2 (100)						
Tragulidae	1 (0)							
Cervidae	2 (40)							
Bovidae	1 (0)	2 (100)						

of relative endemism compared with islands to its east and west. Most families of non-volant mammals on Sulawesi are shared with the Sunda shelf. Even Sulawesian members of the Muridae are related to the Sundanese fauna and not to the New Guinea and Australian murids. However, the endemism of non-volant species on Sulawesi is very high, 97 per cent (61/63). If the Sula islands, which have *Phalanger pelengensis* and *Crocidura* sp. *C*, and the Talaud islands, on which there is *Phalanger ursinus*, are included with Sulawesi then endemism of non-volant species is 100 per cent. The Sulawesian non-volant endemicity is even higher than on New Guinea and Australia, 80 per cent (98/122) and 89 per cent (167/187) respectively.

On the Sunda shelf, 58 per cent (123/213) of the non-volant mammal species are endemic and the figures are no higher for individual islands.

(7). At the genus level Sulawesi also has a high number of endemics (33 per cent, 18/55) compared with the Sunda shelf (19 per cent, 25/182) although lower than the number in New Guinea (37 per cent, 29/76) and Australia (52 per cent 45/88).

The differences between Sulawesi and the other three areas are even more dramatic when only the genera of non-volant mammals are counted, 68 per cent (15/25) endemics contrasted with 22 per cent (20/89), 53 per cent (26/49) and 66 per cent (43/65) for Sunda shelf, New Guinea, and Australia respectively. The high value for Sulawesi reflects the many sciurid and murid genera on the island, compared with the numbers on the Sunda shelf.

The species: origins and antiquity

The ancient character of the endemic Sulawesi non-volant fauna has been recognized both by Groves (1976) and Cranbrook (1981).

The native Sulawesi phalangers are clearly related to the marsupial fauna east of Sulawesi (Feiler 1977, 1978 *a–c*) but the phylogenetic relationships among the three species and between them and other kinds of *Phalanger* are far from clear. Recently, Flannery *et al.* (in press) have proposed an intriguing hypothesis, placing each in a separate genus: *Ailurops* for *P. ursinus*, *Strigocuscus* for *P. celebensis*, with *pelengensis* tentatively retained in *Phalanger*. Furthermore, *Ailurops ursinus* is placed in its own subfamily, and considered to be most primitive of all living phalangers. *Strigocuscus celebensis* is a member of the subfamily Phalangerinae within the tribe Trichosurini; it is the most primitive member of that tribe. *Phalanger pelengensis* is listed in the tribe Phalangerini and retains the most primitive features of that assemblage. *Ailurops ursinus* and *Strigocuscus celebensis* may have reached Sulawesi soon after emergence of land, possibly during middle Miocene times, *Phalangers pelengensis* some time later.

Everyone may not agree with these generic divisions but the interpretations of morphological features, which point to each species as being the most primitive member of its respective group, matches the archaic quality of many mammals native to Sulawesi. Best estimates of the phylogenetic relationships of the two species of anoa, for example, tie them to progenitors known by fossils from the Siwalik beds of India (Groves 1969, 1976). Anoas are not related to any species now living on the Sunda shelf or to the tamarao or dwarf buffalo of Mindoro in the Philippines, with which anoas had been aligned. The tamarao is simply a small-bodied *Bubalus* and is a near relative of the Asiatic water buffalo, *Bubalus (Bubalus) bubalus* (Groves 1969).

The babirusa is very different from other living species of pigs, and from those represented only by fossils. Thenius (1970) placed it in its own subfamily and traced the species as a separate lineage back to Oligocene times. Groves (1981) suggested that the ancestor of the babirusa should be sought in the middle Siwaliks of India and related how some investigators would derive it from *Merycopotamus*, which is an anthracothere whose fossil remains are found in the Siwalik strata. Previously, Groves (1980*a*) had dismissed such a relationship but now speculates that 'really it is impossible to place a finger on any definite resemblance of the Babirusa to any other suid, and perhaps it is an anthracothere after all'.

Sus celebensis, according to Groves (1981), is a remnant of an old line. The species retains many features that are primitive for *Sus* and may have been derived from an ancestral stock living during the early Pliocene. Groves hypothesizes that the *celebensis* lineage was in the Malaysian region and got into Sulawesi before the late Pliocene migration of the *verrucosus-barbatus* lineage to the Sunda shelf.

The primates are also of ancient lineages. The native species of macaque are thought to have been derived from an ancestral population related to the pig-tailed macaque, *M. nemestrina* (Fooden 1969, Groves 1980*b*, Delson 1980). The pig-tailed and Sulawesi macaques are members of the *silenus–sylvanus* group, which has a broad disjunct geographical distribution suggesting to Fooden (1980) early dispersal, especially onto the Sunda shelf, Sulawesi, and the Mentawai islands. According to Albrecht (1978), the short-faced, Sulawesian *Macaca maura* has a cranial conformation that resembles the 'generalized macaque' more closely than any of the other Sulawesian species and may approximate the morphology of the ancestral population on Sulawesi (Groves 1980*b*).

If Niemitz's (1977, 1985*b*) interpretation of the primitive-derived polarities of characters in the Sulawesian tarsiers is correct, then the two species retain more primitive features than do the Philippine and Sundanese species. The origin of the Sulawesi populations, however, are unknown. It has been the view of most primatologists that *Tarsius* is a living fossil but Schwartz (1984) has challenged this conception and noted that after his review of the evidence, 'there are no identifiable fossil tarsoids', and that instead of being truly primitive, *Tarsius* in diet, social behaviour, and many morphological features is very derived. More recently, Simons and Bown (1985) report part of a dentary and toothrow from the early Oligocene strata of Fayum Province, Egypt, which they described as *Afrotarsius* and assigned to the family Tarsiidae. If truly a tarsier, that fragment speaks for the antiquity of the group now represented by the species in Asia of which those native to Sulawesi may be the most primitive.

Shrews are represented by *Crocidura*. Sulawesi, with eight species has more than anywhere else in Asia. Their ancestry is rooted in mainland Asia. Morphological species-limits of each are unambiguous but phylogenetic relationships among them, and between the eight and the *Crocidura* fauna from islands on the Sunda shelf and continental Asia, is unresolved. Despite recent studies by Jenkins (1976, 1982) there is still insufficient information to link the Sulawesian endemics with species occurring to the west. For example, the large-bodied, greyish brown *C. fuliginosa* is found in Indochina and also occurs in Malaya, Borneo, Sumatra, Java, and Timor (Jenkins 1982); in gross external and cranial features, but not in details, *C. rhoditis* and *C. nigripes* resemble *C. fulginosa*. *Crocidura monticola* is small, dark brown, and has been recorded from Borneo, Malaya, and Java; in body size, colour and cranial configuration, the Sulawesian *C. levicula* is similar to *C. monticola*. The other Sulawesian species bear no close morphological relationship to any other species of Asian *Crocidura*. Lately, I have been studying Asian shrews to determine whether the Sulawesian species represent several dispersal events or are the evolutionary outcome of speciation from one ancestral stock. Perhaps the fauna resulted from a mixture of the two phenomena. That possibly all or at least most of the species-diversity on Sulawesi reflects insular differentiation may suggest a long history of *Crocidura* on Sulawesi.

Sulawesian squirrels certainly had their ancestral beginnings in mainland Asia. Unfortunately, phylogenetic relationships are still unclear. *Prosciurillus* resembles certain species of *Sundasciurus*, but *Rubrisciurus* and *Hyosciurus* cannot be tied to other genera. The degree of morphological distinctness characteristic of these latter two points to long isolation on Sulawesi and ancestry in an old fauna that no longer exists on the Sunda shelf.

Although phylogenetic studies of murids

native to the Indo–Australian region are incomplete, my inquiries indicate that the fauna on Sulawesi is most closely associated with that on the Asian mainland rather than with the endemic faunas peculiar to the Philippines, New Guinea and Australia, and the Lesser Sunda Islands. Furthermore, the endemic Sulawesi rats and mice also demonstrate an archaic nature. The 36 species form nine generic clusters.

(1). *Crunomys*, which is also part of the Philippine fauna, is characterized by some derivations but generally possesses a suite of primitive features (Musser 1982). There is nothing like it on the Asian mainland, either living or fossil.

(2). *Echiothrix*, the large-bodied shrew rat, is highly specialized for living on the forest floor and eating earthworms. Its obvious derivations include body form and cranial conformation (Musser 1969) but its dental characteristics are primitive. *Echiothrix* is not related to either the shrew rats in the Philippines or to the shrew-like mice of New Guinea. It has no living relative on the Asian mainland.

(3). The smaller-bodied shrew rats, *Tateomys* and *Melasmothrix*, are unique (Musser 1982). They, like *Echiothrix*, are also characterized by a suite of morphological specializations reflecting forest life and a worm and insect diet. Their dentitions, however, have retained many primitive features. Aspects of their morphologies recall species in the genus *Mus*, but the association reflects shared archaic murid characteristics. Neither *Tateomys* nor *Melasmothrix* are closely related to *Mus volcani* from Java and *M. crocidoides* of Sumatra, the only *Mus* known to be native on the Sunda shelf.

(4). The undescribed genus listed in Table 7.3 has a primitive dentition and is unlike any other murid genus. It has no close living relative and cannot be tied to any known fossils.

(5). *Margartamys*, *Lenomys*, and *Eropeplus* are each characterized by conspicuous diagnostic characters but all are probably more closely related to one another than to any other cluster of species on Sulawesi. In turn, the three share some significant derived features with *Lenothrix* on the Sunda shelf. *Lenothrix* has definite dental specializations but in overall morphology is one of the most primitive rats now found in Sundaland (Musser and Newcomb 1983). A reasonable hypothesis is that *Lenothrix* represents a Sundanese relict derived from the ancestral lineage out of which the Sulawesian genera evolved.

(6). *Hearomys* is also found on Borneo and Palawan. Except for its very long tail relative to head and body, and a very few other morphological specializations, the Bornean and Palawan members of *Hearomys* are the most primitive of the murid fauna native to those islands (Musser and Newcomb 1983). Dental characteristics of the Sulawesian species indicate that they have retained more primitive features than their Sundanese relatives.

(7). *Maxomys* is primarily Sundaic but also gets into Indochina (Musser and Newcomb 1983). The species on Sulawesi form two clusters: *M. muschenbroekii* and *M.* sp. *C* are one; the other consists of *M. hellwaldii*, *M. dollmani*, and *M.* spp. *A* and *B*. Whether all six were derived from a single lineage or whether each group can be tied to different species on the Sunda shelf is still under study.

(8). Species of *Bunomys*, *Taeromys*, and *Paruromys* were once included within *Rattus* but the diagnostic morphological features of each are uncharacteristic of that genus (Musser and Newcomb 1983). Their origins definitely lie to the west where their approximate morphological counterparts may be *Sundamys* and *Berlymys*.

(9). Native *Rattus* on Sulawesi is represented by *R. hoffmanni*, *R. marmosurus*, *R. xanthurus*, and an undescribed species from Pulau Peleng (Table 7.3). Their origins are with the fauna on the Sunda shelf and not with clusters of *Rattus* native to the Sula islands, Moluccas, New Guinea, or Australia (Musser and Newcomb, 1983; Musser and Heaney 1985).

The combination of primitive and unique characters, and the spectacular nature of the morphological specializations in many of the endemic Sulawesian murids suggest that much of the fauna evolved in isolation on Sulawesi for a long time and had its phylogenetic roots in an ancient group on the Asian mainland that is now gone except for some relictual species, such as those in *Lenothrix*, *Haeromys*, and *Maxomys*.

Relationships of the bats native to Sulawesi are ambiguous. Three genera and nine species are endemic, which is only 15 per cent of the total bat fauna. Whether these endemics were derived from primitive ancestral stocks, and what their phylogenetic relationships are to the rest of the Sulawesian bats and to those occurring outside that region, are unclear because analysis of phylogenetic relationships among Indo–Australian species of bats are unavailable. About all that can be said is that most of the Sulawesian bat fauna is a blend of species, with some occurring in various combinations nearby.

The origin of the entire bat fauna east of the Sunda shelf was originally in the west, considering the low endemicity of the fauna east of the Sunda shelf and the fossil record of bats in that region. The earliest fossils are from the Miocene of Australia. To Koopman (1984), this 'makes any hypothesis that some part of the Australian bat fauna was there when Australia commenced its northward drift in the Eocene (or reached there across broad expanses of open ocean afterward) highly improbable. Rather a percolation of numerous species across a series of relatively narrow water gaps from the north and west, starting in the Miocene, seems indicated.'

OVERWATER DISPERSAL, LAND BRIDGES, OR RAFTING

To me, Sulawesi is a large oceanic island supporting a native and endemic mammal fauna that, with the exception of phalangers, had its origin on mainland Asia. I agree with Cranbrook (1981) that 'Cumulatively the evidence of all vertebrate groups strongly suggests that since its (most recent) emergence there has been no direct, unbroken subaerial connection between Celebes and the principal landmasses of the Sunda shelf. All immigrants from Sundaland have probably been obliged to cross a sea barrier. Notwithstanding geological evidence of Celebes dual origin, its Laurasian segment has apparently not served as a raft carrying with it any strictly terrestrial (including freshwater) vertebrate of Asian origin. In their present composition, the vertebrate faunas provide no evidence to support the suggestion that the Makassar Strait was closed to a period during the late Pliocene. . . . Moreover, any Quaternary land connection formed between Borneo and Celebes as a consequence of lowered sea-level appears likely to have been incomplete or short-lived.'

Biologists (Groves 1976; Meijer 1982), geologists (Audley-Charles 1981; Sartono 1973) and a combination of the two (Audley-Charles and Hooijer 1973) have proposed either land connections between Borneo and Sulawesi, or between Java and Sulawesi, or closure of the Makassar Strait with western Sulawesi abutting eastern Borneo (Katili 1978) at times during the Miocene, Pliocene, or Pleistocene. The composition of the Sulawesian mammal fauna simply does not support the hypotheses of either land bridges or close juxtaposition of Borneo and Sulawesi.

Fifteen mammal families of which six are bats are represented on Sulawesi. Of these, the Phalangeridae is clearly related to the New Guinea and Australian marsupial faunas. Phalangers are among the few marsupial groups found on single islands and archipelagos outside mainland New Guinea and Australia (Tables 7.6 and 7.7), which suggests that at least some of the species are better suited to range expansion through dispersal events than are species in most of the other marsupial families.

The other eight families of non-volant mammals are mainland Asian in affinities. Out of the 28 families of non-volant mammals found on the Sunda shelf, 20 are not represented on Sulawesi; some, such as the Talpidae, Rhizomyidae, and Leporidae, do not occur closer to Sulawesi than either the Malay Peninsula or Sumatra. But even Borneo, the nearest Sunda island to Sulawesi, has 16 families not represented on Sulawesi. If land connections between the Sunda shelf and Sulawesi existed during the period from middle Miocene to late Pleistocene, why are more Sundanese families not now part of the Sulawesi fauna? Where are the tree shrews, for example; where are the mustelids, cats, canids, other civets, and the mongooses; where are the larger mammals such as deer, rhinoceroses, and tapirs? Why are there not more genera of primates on Sulawesi, why no pongids? Why so few species of squirrels and no flying squirrels on Sulawesi as compared with the Sunda shelf, and especially with Borneo which has 35 native squirrel species.

Contrasted with Borneo in particular and the Sunda shelf in general, the fauna of Sulawesi is depauperate and unbalanced. Forty per cent of it is made up of bats (34 per cent of the Sundaic fauna consists of bats), and a third is composed of rats and mice (15 per cent of the total fauna on the Sunda shelf are murids). Furthermore, nearly all the non-volant species are endemic and endemicity is also high at the generic level. The Sulawesian picture is characteristic of an unbalanced island fauna that originated by a sweepstake route over a large water barrier (Dermitzakis and Sondaar 1979).

Sulawesi is the first large oceanic island located east of Borneo and is separated from the Asian continental shelf by the deep Makassar Strait, which has been open and a marine sedimentation basin possibly since Cretaceous and certainly since Eocene times (see Audley-Charles 1981, and references cited there). The modern native fauna of Sulawesi is richer in species and genera than is usual for most other oceanic islands because its source area was and is rich in species, genera, and families; because Sulawesi is now an extensive land surface with great topographical relief and even in the past was probably an archipelago of large islands; and because adaptive radiation in several groups has produced clusters of endemics. The fauna is balanced within the physical and ecological boundaries of the island but unbalanced in its diversity of major groups compared with the Sundaic fauna.

Migration to Sulawesi may have begun during late Miocene or early Pliocene times for apparently what is now eastern Sulawesi was subaerial in middle Miocene and part of western Sulawesi was above sea level by late Miocene (Audley-Charles 1981). The degree of morphological diversity and endemism characteristic of the non-volant mammals suggests that many got to the Sulawesi region early in its geological history, most from the Sunda shelf, and a few from the east.

If the bear phalanger is as primitive and from as old a lineage as postulated by Flannery *et al.* (in press), it is tempting to explain its presence by rafting on the Gondwanic segments which moved westward in the Miocene to form eastern Sulawesi. No data however, support such an hypothesis and if rafting was involved (Audley-Charles 1981), why are not other kinds of marsupials living now on Sulawesi and islands between there and the Moluccas? (I do not accept the supposition that a diverse marsupial fauna once existed on Sulawesi but lost out in competition with placentals from the Sunda shelf.)

Finally, no evidence supports the hypothesis that Sulawesi was a centre of origin for mammals now living in other regions as Groves (1976, 1980*b*) suggested for *Tarsius* and *Rattus*. The Philippine and Sundaic tarsiers may or may not have been derived from the Sulawesian species. Perhaps they evolved from a primitive ancestor that was living on Sundaland at about the time tarsiers migrated to Sulawesi. Populations on Sulawesi retained primitive features, those elsewhere became specialized. In the absence of fossils, this hypothesis is as valid as the other. *Rattus* is not so diverse on Sulawesi as was thought, and lineages of the species probably go back to ancestral populations on the Sunda shelf.

THE PLEISTOCENE FAUNA

All the Pleistocene vertebrates recorded from Sulawesi are represented by specimens collected in either river terraces or strata exposed along the Sungai Walanae in the south-western arm in the region below Danau Tempe (see the map in

Barstra 1977). Sharks and sting rays; a crocodile tentatively identified as *Crocodylus siamensis*, which still lives on the Sunda shelf; the giant land tortoise, *Geochelone atlas*; an elephant, *Elephas celebensis*; a pygmy stegodont, *Stegodon sompoensis*, and the larger stegodont, *S. trigonocephalus*; an anoa, *Bubalus depressicornis*; and two kinds of pig, *Celebochoerus heekereni* and *Sus celebensis* (Hooijer 1969, 1972, 1975) comprise the assemblage. Like the anoa and *Sus celebensis*, the probiscidians and *Celebochoerus* represent an old fauna related to species known by fossils from Pliocene beds of the Siwaliks in India and middle to earlier Pleistocene strata on Java (Thenius 1970; Hooijer 1975; Groves 1976).

Controversy surrounds the fossils. Stone artefacts (the Cabenge industry) were collected on the Walanae terraces along with some of the fossils and both fauna and industry have been assumed to be contemporaneous. Barstra (1977) has reviewed the evidence in a report of the stratigraphy of the south-western peninsula. He pointed out that the fossils came from two layers.

A controversy of identification is associated with the tortoise pieces and some of the mammal fragments. The proboscideans have been identified by Hooijer as *Elephas celebensis*, which is also known from Java; *Stegodon sompoensis*, which he also recorded from Flores and Timor; and *Stegodon trigonocephalus*, with specimens also from Java, Flores, and Timor (Hooijer 1975). A fossil *Stegodon* has also been found in Sumba, but its specific identity is clouded (Hooijer 1981). The giant land tortoise was identified as *Geochelone atlas* and, according to Hooijer (1971), not only lived on Sulawesi but also Timor and Java, and its remains are known from the early Pleistocene Siwalik deposits of India. Not so, contended Sondaar (1981), who argued that each sample of *Geochelone* from Java, Sulawesi, and Timor represented a different species, as did each sample of the pygmy stegodont. Hooijer (1982*a*) did not agree and declared that there was no reason to revise his earlier conclusions about *Geochelone* and *Stegodon*. Recently, Sondaar (1984) has even argued that the Javan fossils described as *Elephas planifrons* and *E. celebensis* are really examples of the Javan pygmy stegodont, *Stegodon hypsilophus*, and that the remains identified as *Elephas celebensis* from Sulawesi is not *Elephas* at all.

Another question about the fossils are their geological age. Barstra (1977) noted that the upper levels of the Walanae formation cannot be younger than early Pleistocene and Sartono (1979) claims they are upper Pliocene. Fossils actually found in those layers would be from either of those periods but what of the fragments that were discovered in the terrace fills? Some, as Barstra indicated, were stream-eroded and redeposited in the terrace sediments. What about the others that were found on the surface for which there is no provenance information?

I have not studied the fossils at first hand and so cannot give my identifications. Clearly the samples should be examined again to test the opposing hypotheses of identifications presented by Hooijer and Sondaar. It is significant to me that during Pleistocene times, the land vertebrates of the region now known as south-western Sulawesi sustained populations of a giant land tortoise, a large stegodont, a pygmy stegodont, another kind of pygmy probiscidian, two kinds of pigs, and an anoa. We do not know if all lived together; even Hooijer (1975) declared that 'although *Elephas*, *Celebrochoerus* and *Stegodon* are found together we do not have sufficient stratigraphic evidence to prove contemporaneity'. This is especially true for the anoa and *Sus celebensis*. Are the fossil fragments of these species really from the same sediments and did they live during the same time as the *Geochelone–Stegodon–Celebochoerus* fauna or are they from much younger deposits and from a later time when south-western Sulawesi was populated by the fauna we know today?

I puzzled over these questions as I worked in the primary forests of central Sulawesi. Why were no probiscidians and *Celebochoerus* living there? After I learned something about the forest and fauna, I was impressed with the numbers of species encountered, with the range of habitats they were found in, and with their different morphological adaptations. Untrammeled central Sulawesi had a sense about it of balanced communities, integrated faunas and floras; there seemed no room for another pig or any proboscidian. What then was so peculiar about the Pleistocene fauna? Was it really part of Sulawesi?

Back in the Museum, I sought records of Pleistocene fossils from elsewhere in Sulawesi. None existed. Why? Were they not preserved? Do they exist but have not yet been discovered? Or could it be that during the Pleistocene when the *Geochelone–Stegodon–Celebochoerus* fauna existed, the region that is now south-western Sulawesi was a large island; that its fauna of large mammals (we have no idea what small mammals may have existed at the time) had nothing to do with the mammals living in the main Sulawesi area to the north; and that the tortoises and stegodonts were instead related to tortoises and stegodonts on an island to the west in what is now Java and land to the east in what is now called the Lesser Sunda Islands?

This is why I have not included the proboscidians and *Celebochoerus* in my review of the mammals native and endemic to Sulawesi. And I conclude by putting forth the hypothesis that the Pliocene (or early Pleistocene) proboscidians and *Celebochoerus* were part of an island fauna, unbalanced, free of major predators, and unrelated to the real Sulawesi fauna; that the Pleistocene examples of *Bubalus depressicornis* and *Sus celebensis* are from much later sediments and the animals lived at a later time. Such a reconstruction suggests that extinction of the stegodont fauna may have been due to competition from the Sulawesi fauna living in the north when it moved south after the south-western island was joined to the central core. Probably changes in climate and vegetation were also involved, and even very early interaction with humans, as Glover (1981) suggested.

The speculation that south-western Sulawesi was once an island is not novel. It formed part of the ancient Sulawesi archipelago in Fooden's (1969) reconstruction and was integral to explaining speciation events among populations of macaques. Recently, Audley-Charles noted (1981) that the flora of Gunung Lompobattang is more closely related to that found in the Lesser Sunda Islands than to floras of western Sulawesi, Java, or Borneo. In answer to why such a flora occurs on Gunung Lompobattang and not in other parts of south-western Sulawesi, he pointed out that the mountain is composed of 'Miocene, Pliocene, and Quaternary volcanics . . . and may have been an isolated island in western Celebes when most of that territory was covered by the sea in the late Miocene and early Pliocene'. He explained that the flora may have been derived from eastern Sulawesi and been isolated by the migration of Bornean flora to western Sulawesi.

Results of careful fossil collecting under tight stratigraphical control in south-western Sulawesi and other regions of the island where Pliocene and Pleistocene strata are exposed will provide information to either support or falsify the hypothesis. Careful review of all the fossils now in museums will clarify phylogenetic relationships among the stegodont faunas and may help us understand the nature of either past insular isolation or continuous land surface from Java to Timor. Whatever the answers, we will be enriched by learning more about the diversity and evolution of the mammalian fauna in that part of the world east of Borneo and Bali.

8 A PLANT GEOGRAPHICAL ANALYSIS OF SULAWESI

M. M. J. van Balgooy

The geographical relationships are examined by two different numerical techniques of the 4222 species of 540 genera of flowering plants growing in Malesia for which there is a modern taxonomic treatment.

At both species and genus level the strongest affinities of the Sulawesi flora are to the north (Philippines), east (Moluccas) and south (Lesser Sunda Islands). Most plants appear to have entered from one of these directions. There is much less affinity with Borneo, lying to the west across the Makassar Strait, although Borneo is very close. Wallace's line, which runs down the Strait, is confirmed to be an important phytogeographical boundary. Floristic disjunctions between different parts of Sulawesi are less than with neighbouring islands. The island is a coherent unit.

INTRODUCTION

Wallace's line has always received more attention from zoologists than from botanists. Yet the importance of the line and the critical position of Sulawesi have long been recognized by botanists as well.

In his well-known study of Malesian mountain plants van Steenis (1934–36) concluded that these plants have entered the archipelago along three routes, all of them involving Sulawesi: the Sumatra track, the Luzon track and the New Guinea track (see Fig. 8.1).

Lam (1945) in an analysis of 734 species of Sulawesi plants found that the island had more species in common with the islands to its west than those to its east. He regarded Sulawesi as belonging floristically to the Sunda shelf. The distribution of his species suggested that very few have reached Sulawesi by directly crossing Makassar Strait. He concluded that migration had been primarily by way of a northern route, and less so via an eastern and a southern route (see Fig. 8.2). Thus he recognized the importance of Wallace's line, or rather the importance of Makassar Strait, as a barrier to dispersal and migration.

Van Steenis (1950) established the limits of Malesia, the area covered by the Flora Malesiana project, by establishing the complete distribution of all genera occurring in the Malay archipelago. He found three places with a very high accumulation of generic distribution limits (Fig. 8.3). These places, which are not crossed by many genera in either direction, he called 'demarcation knots'. They are situated at the Isthmus of Kra, between Luzon and Taiwan, and between New Guinea and north Australia. These 'knots' were considered to mark the borders of Malesia. By following the same procedure an inner division of Malesia was proposed into west Malesia (Sumatra, Malaya, Borneo, and the Philippines), south Malesia (Java and Lesser Sunda Islands), and east Malesia (New Guinea, Moluccas, and Sulawesi), thus recognizing Wallace's line, at least between Borneo and Sulawesi.

Hamilton (1979), Audley-Charles (1981 and this volume, Chapter 2), and others have dealt with geological reconstructions of the Malesian region in the light of plate tectonics. The present analysis was made to see if the distributions of plants reveal anything about the origin and history of the Sulawesi flora.

METHOD

The basis of this paper is a study to be published in full elsewhere (de Koning and Sosef *in prep.*) of the 4222 species and 540 genera of flowering

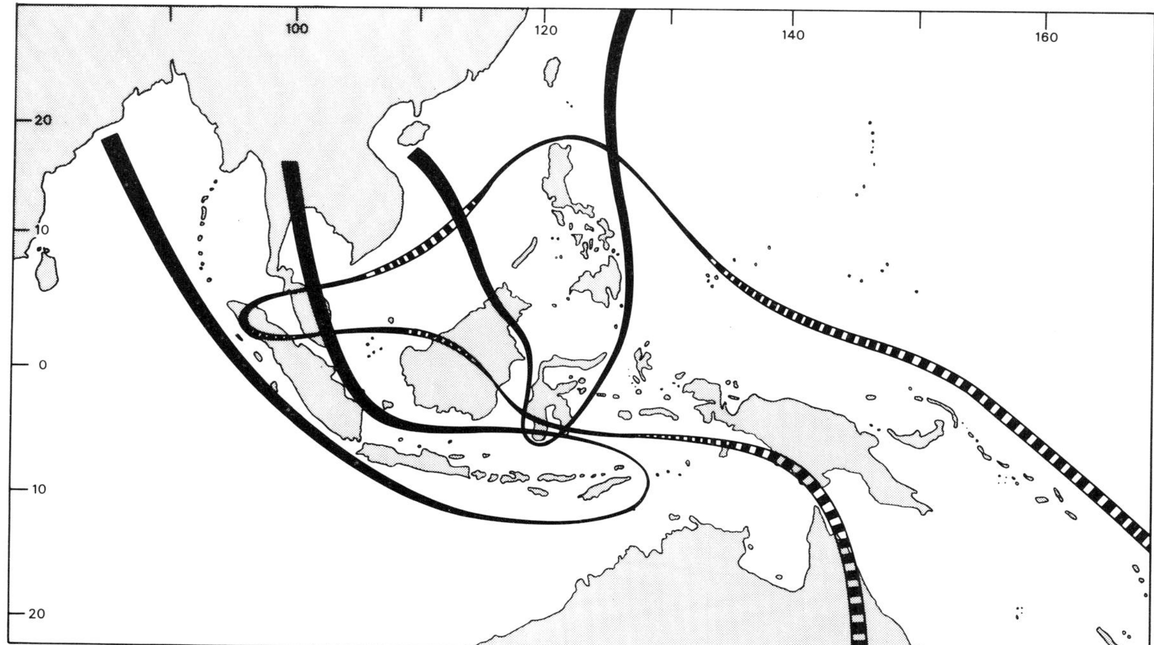

Fig. 8.1. The migration tracks of van Steenis (van Steenis 1965, Fig. 11). These show entry of mountain plants to Malesia from north-west, north, and south-east: respectively the Sumatran, Luzon and New Guinea tracks.

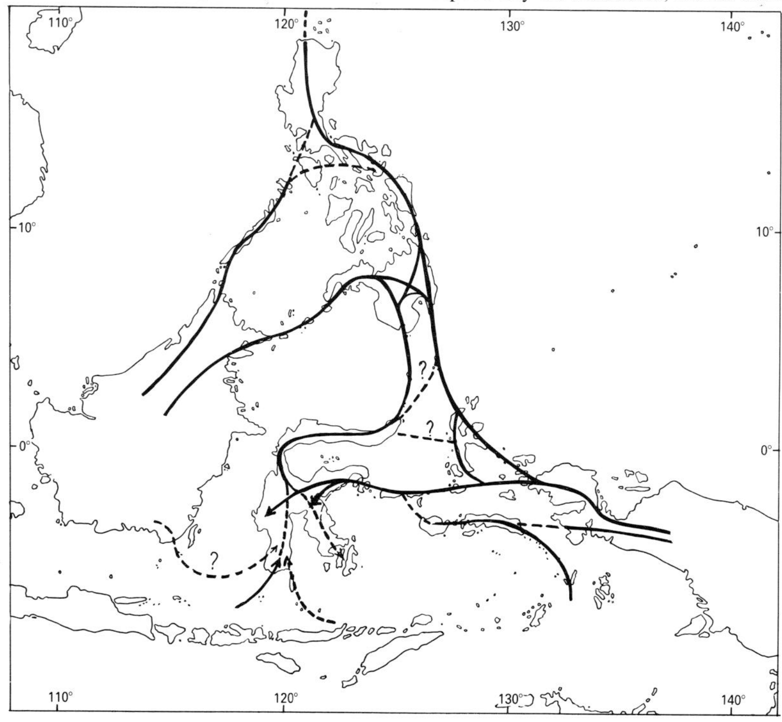

Fig. 8.2. Migration tracks of plants to Sulawesi according to Lam (1945).

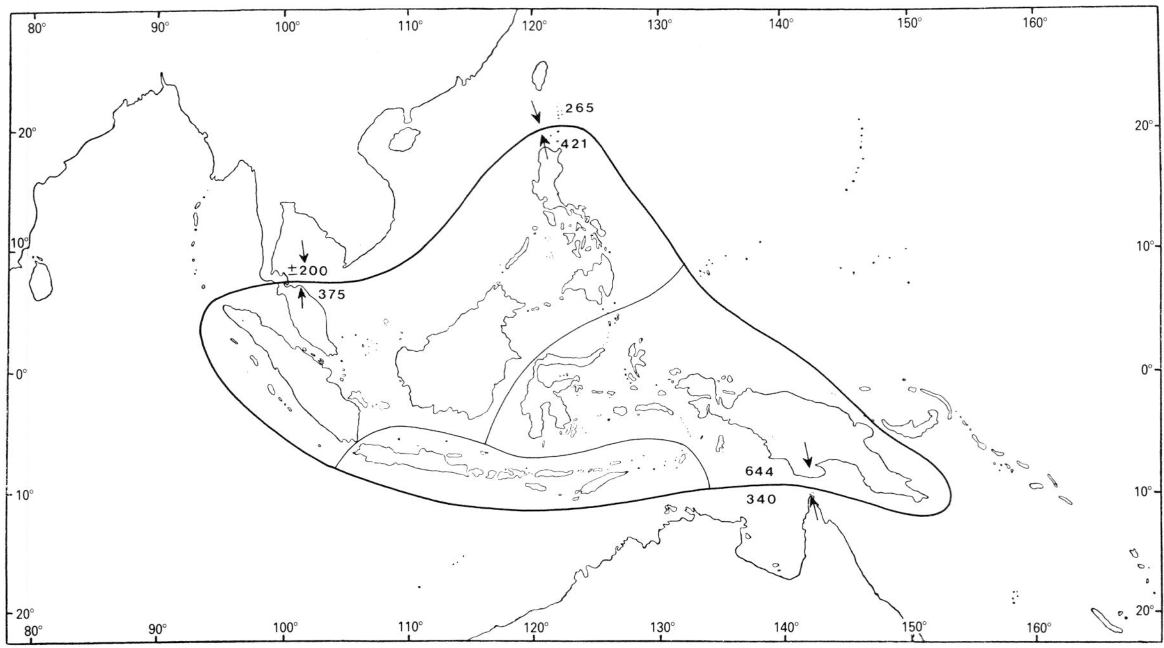

Fig. 8.3. The limits and subdivisions of Malesia according to van Steenis. The figures show the number of genera that do not cross the three main demarcation knots.

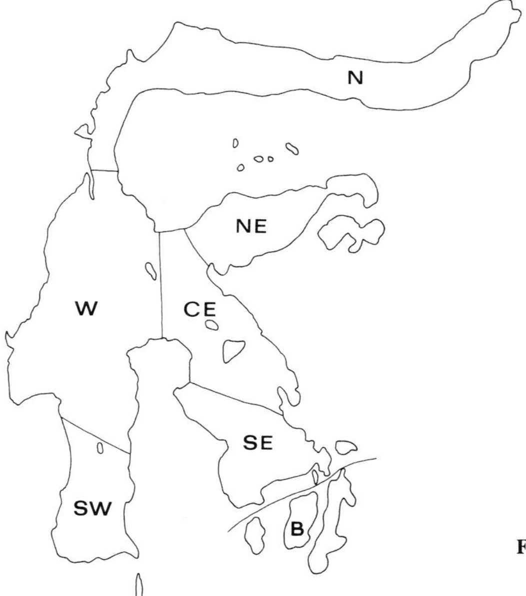

Fig. 8.4. The seven parts of Sulawesi accepted in this study: N = north arm; NE = north-east arm; W = west part, CE = central east part; SE = southern part of south-east arm; B = Buton, Muna and Kabaena, SW = south-west arm.

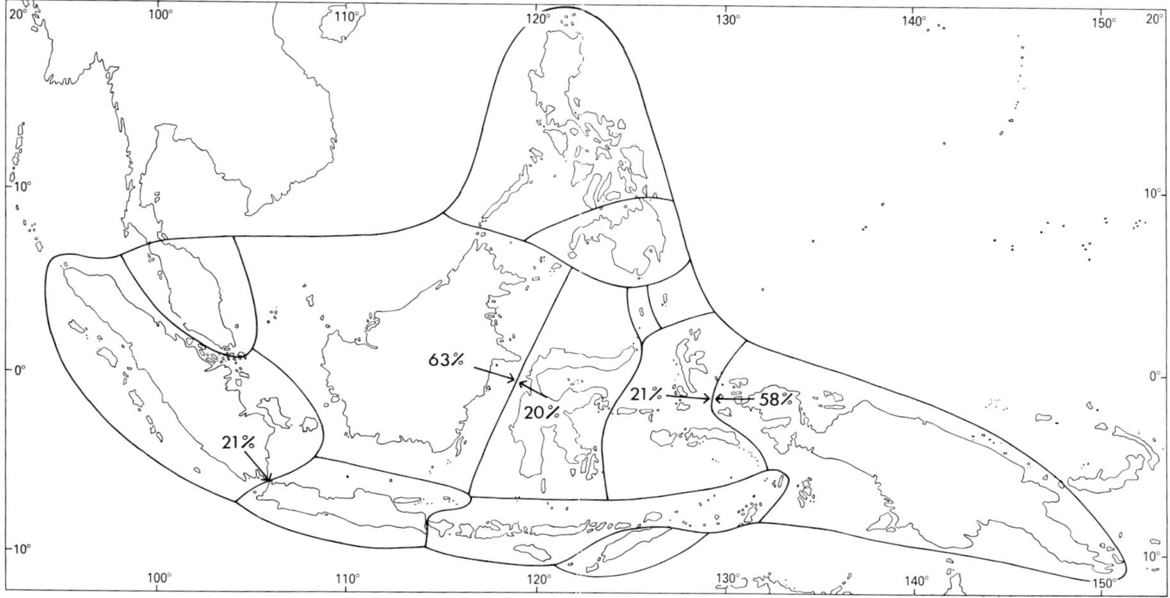

Fig. 8.5. The unit areas of Malesia used in this study. The figures show the percentages of all species that do not cross certain boundaries (fuller details in text).

plant families treated in volumes 4–10(1) of *Flora Malesiana* series I (1950–84), supplemented, where possible, by more data that have become available after completion of the *Flora Malesiana* revisions. As regards Sulawesi, special mention must be made of the collections by W. Meijer in 1975 and 1976, and of the Dutch-Indonesian expedition in 1979.

For the purpose of this analysis Sulawesi was divided into seven parts (Fig. 8.4). The surrounding areas were also broken up into units (Fig. 8.5), and the distribution over these units was tabulated for all the species and genera. Furthermore, data were assembled for all species on habit, ecology and dispersal. *Flora Malesiana* provides the most recent, reliable and comparable information. Using this distribution matrix it was possible to calculate the floristic affinity of each part of Sulawesi as well as the floristic affinity of the island as a whole to each of the surrounding units or combinations of them. We also calculated the affinities of certain groups of taxa, for example only the forest plants and only the mountain plants.

Two different tests for floristic affinity were made. Firstly, the formula of Kroeber (1916) was used. The preference for this formula over other more sophisticated formulae has been explained in my analysis of generic distributions in the Pacific (van Balgooy 1971). The results are shown in Tables 8.1 and 8.2. Secondly, cluster analysis was carried out to produce a number of dendrograms (Figs 8.6–8.8). For this analysis the 'group average' method was preferred over 'simple linkage' (de Koning and Sosef *in prep.*). More about these methods can be found in Sneath and Sokal (1973).

The strength of certain demarcation lines within Malesia was also tested by calculating the number and percentage of taxa that do not cross such a line in one way or the other. The relative importance of migration tracks that are suggested on the basis of distribution patterns was also investigated.

Table 8.1

Floristic affinities of Sulawesi at species level, expressed in Kroeber's coefficient. Island units are shown in Fig. 8.5

(a) *Lowest grouping*		(b) *Middle grouping*	
LSI less Timor	57.9	Moluccas	60.3
Seram + Buru	56.6	LSI (including Timor)	56.6
Java	56.3	Java	56.3
N. Philippines	52.6	Philippines	52.6
Halmahera	52.6	Sumatra	46.4
Sula islands	51.7	New Guinea	42.8
Timor	49.9	Borneo	42.3
Talaud	49.6	Malaya	39.5
S. Philippines	48.5		
Sumatra	46.4	(c) *Highest grouping*	
New Guinea	42.8	Moluccas	60.3
Borneo	42.3	Java + LSI	58.1
Malaya	39.5	Sunda block*	45.7
		New Guinea	42.8

* Sumatra + Malaya + Borneo
LSI = Lesser Sunda Islands

Table 8.2

Floristic affinities of Sulawesi at genus level, expressed in Kroeber's coefficient.

(a) *Lowest grouping*		(b) *Middle grouping*	
N. Philippines	81.0	LSI (including Timor)	83.9
Java	80.6	Philippines	81.8
New Guinea	79.9	Moluccas	81.8
Borneo	78.5	Java	80.6
LSI less Timor	77.6	New Guinea	79.9
S. Philippines	77.2	Borneo	78.5
Sumatra	77.1	Sumatra	77.1
Seram + Buru	74.9	Malaya	72.9
Malaya	72.9		
Halmahera	72.0	(c) *Highest grouping*	
Timor	67.5	Philippines	81.8
Sula islands	63.2	Moluccas	81.8
Talaud	62.0	Java + LSI	80.9
		New Guinea	79.9
		Sunda block	78.1

LSI = Lesser Sunda Islands

The complete flowering plant flora of Malesia comprises around 2300 genera and *c.*30 000 species. This study thus deals with *c.*25 per cent of the genera and *c.*15 per cent of the species. Some caution is therefore needed when interpreting the results:

(a) Although the sample size is fairly large, it is not really a random sample. In *Flora Malesiana* there is a bias towards smaller families. Most of the large families have not been treated.

(b) It would be naive to expect that all treatments are equal. There are obvious differences in species concept.

(c) Certain areas that are of critical importance to this analysis are appallingly under-collected; to name a few: the lowland flora of east Kalimantan, the Moluccas, Irian Jaya, and within Sulawesi, the east arms.

(d) Distributions are dynamic in time and it is hazardous to use static ditribution data to interpret dynamic processes in the past.

(e) Numerical data can give misleading information. For example *Xanthophyllum* (Polygalaceae), a genus of trees with 93 species mostly restricted to Malesia, has always been regarded as Indo-Malesian with only two outlying immigrants in Queensland. Detailed analysis, following close taxonomic study, has recently led, however, to the conclusion that the genus is in fact most likely of Australian origin and that it has undergone secondary explosive radiation in west Malesia (van der Meijden 1982).

RESULTS

Floristic affinity

Sulawesi has more species in common with areas to its west than to its east. Based simply on these figures one would conclude that the flora is more allied to that of the Sunda shelf than to that of the Sahul shelf. More informative results are obtained when calculating floristic affinities of Sulawesi using Kroeber's formula (see Table 8.1) or group averaging (see Fig. 8.6). At species level the affinity of Sulawesi is clearly strongest with the Moluccas, Lesser Sunda Islands + Java, and the Philippines. Sulawesi is floristically rather remote from both the two largest islands of the Sunda and Sahul shelves (Borneo and New Guinea). Keeping all units apart Table 8.1a shows highest Kroeber figures for the Lesser

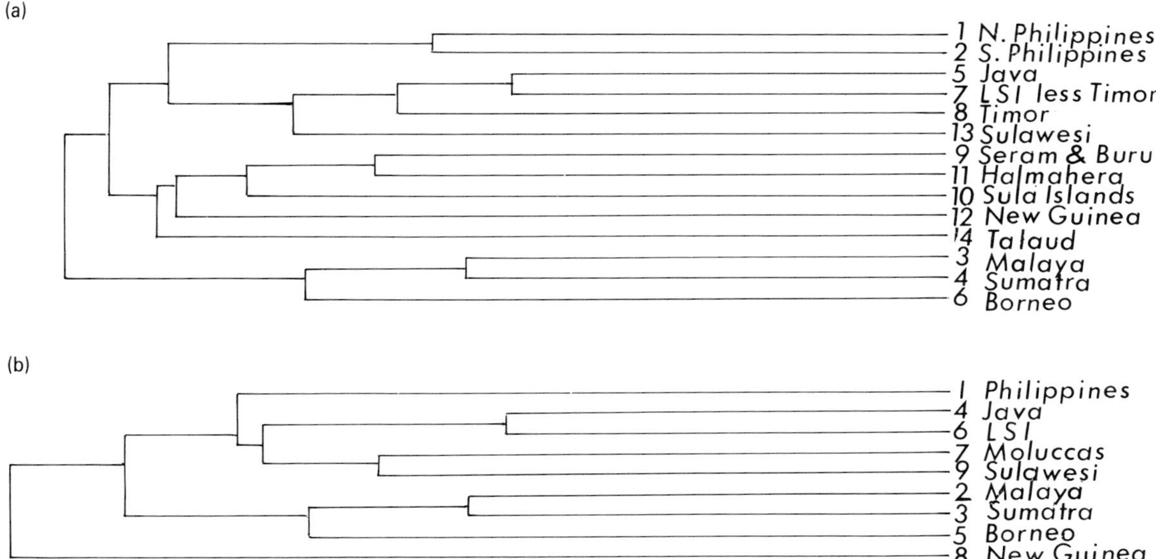

Fig. 8.6. Dendrogram showing the affinity of Sulawesi at species level (including endemics) according to the group averaging method; (a) individual units (see Fig. 8.5); (b) groups of units.

Sunda Islands (LSI) less Timor, Seram and Buru, Java and north Philippines. When grouping together closely allied units (Table 8.1b) the affinity is largest with the Moluccas, followed by Java and LSI, and then the Philippines. According to the group averaging method, Sulawesi groups together with the LSI + Java first, and these together with the Philippines.

Similar procedures can be followed using the genera (Table 8.2 and Fig. 8.7). At genus level the highest Kroeber figures are found for north Philippines, Java and New Guinea (Table 8.2a). Again, one can take together certain units, Table 8.2b, in which case the highest values are found for the LSI, the Philippines, and the Moluccas.

According to the group averaging method Sulawesi groups together with New Guinea, and these two group together with the Moluccas (Fig. 8.7a). When applying group averaging to larger island blocks (Fig. 8.7b), Sulawesi groups together first with the Philippines; these together with the Moluccas; and the three together with New Guinea.

The results so far are based on all species in the study including the endemics. The phytogeographical status of an area is determined by its proportion of endemic taxa, but they also obscure the relations with other areas. For this reason we have calculated group average coefficients leaving out the endemics; Fig. 8.8 shows that at species level Sulawesi groups together first with the Moluccas, and these together with New Guinea. At genus level Sulawesi groups together first with the Philippines, and the two together with the Moluccas and New Guinea. At genus level there is little difference from the analysis including the endemic genera (Figs 8.7, 8.8b).

Turning to floristic affinities between the various parts of Sulawesi, it appears that the closest affinity of each part is always with another part of the island. In other words Sulawesi forms a closed phytogeographical unit. Of the 297 genera none is confined to Sulawesi, but of the 729 species 100 are endemic to Sulawesi. These endemics are best represented in the west and central-east parts. We did not find stronger affinities between the various western parts than between them and the eastern parts. Nor did we

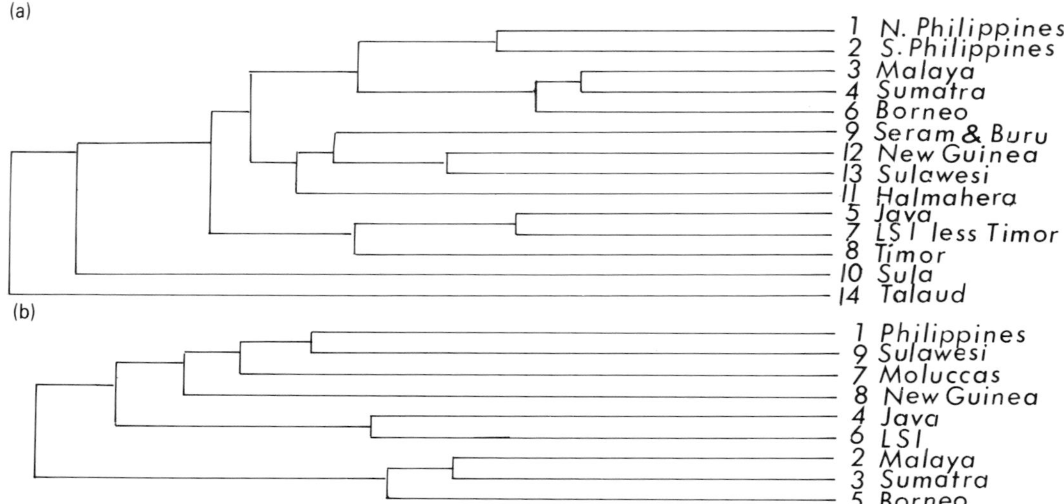

Fig. 8.7. Same as Fig. 8.6 at genus level.

find strikingly high affinities between the eastern parts and island groups of east Malesia. We did find, however, a strong affinity between the north arm and the Philippines, and between the south-west arm and the Lesser Sunda Islands. Van Steenis (1972, 1979) had already pointed out the strong affinity of Mt. Lompobattang with the LSI.

In summary, the Sulawesi flora appears to be most distinctly allied to the flora of the Philippines, the Moluccas, and the LSI.

These same relations can be found in the various categories into which the flora can be subdivided, e.g. when considering only the rain forest flora. The affinity of the monsoon flora of Sulawesi with areas with monsoon climate is even stronger than the affinity of the flora as a whole. The trees and climbers of the Sulawesi flora show strongest affinity with those of the Moluccas and the shrubs and herbs show strong alliance to those of the Lesser Sunda Islands and Java. The flora of the ultrabasic soils shows a relatively high affinity with New Guinea.

Demarcation lines

We have calculated the numbers of taxa in various parts of Malesia which do not occur beyond a certain line, e.g. taxa occurring in Borneo and not in Malaya, taxa occurring in Borneo not in Sulawesi, taxa in New Guinea not occurring in the Moluccas, etc. It appears that there are certain strong demarcation lines. We will not discuss the demarcation lines of Malesia towards outside areas because we are unable to produce reciprocal figures, e.g. many taxa in Luzon do not extend to Taiwan, hence there is a very strong demarcation between Luzon and Taiwan, but we do not know how many taxa occurring in Taiwan do not extend to Luzon.

Inside Malesia there is a very strong demarcation between Borneo and Sulawesi at species level (see Fig. 8.5). We found that 456 species occurring in Borneo do not occur in Sulawesi or east of this island (48.3 per cent); including the endemics the figure rises to 63 per cent. The demarcation line between Borneo and Sulawesi, however, is rather one-sided. If we look at the reciprocal figures, 52 species of Sulawesi do not occur west of the Makassar Strait, or 152 if we include the endemics (i.e. 20 per cent of the total).

For comparison 132 species (or 259 including endemics, viz. 21 per cent) of Sumatra do not occur in Java or east of Java.

Another strong demarcation line is found westwards of New Guinea: 21.9 per cent of the non-endemic species and 58.1 per cent of all New

(a)

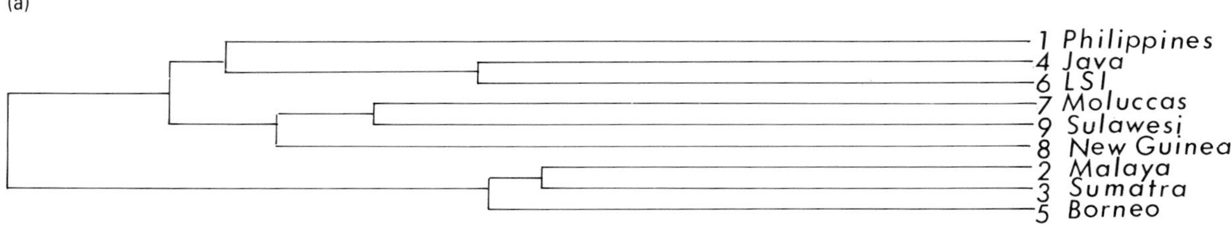

(b)

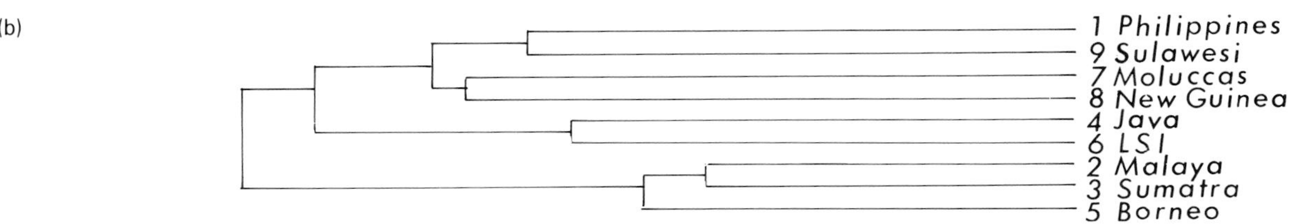

Fig. 8.8. Group averaging at (a) species and (b) genus level excluding endemics.

Guinea species do not occur west of the Vogelkop peninsula. The reciprocal figures are respectively 17.3 and 21.2 per cent.

At genus level we find lower figures, as is to be expected. Borneo–Sulawesi: 16.1 per cent (17.2 per cent), reciprocal figures; 3 (3) per cent. For New Guinea–Moluccas we find 9.7 (11) per cent and 4.9 (4.9) per cent.

The demarcation of Sulawesi against the Moluccas, the LSI, and the Philippines is weak. These results confirm the conclusions reached above on floristic affinity, viz. that Sulawesi is closest to these island groups.

Migration tracks

The distributions of many species in Sulawesi suggest that they have reached the island by using a limited number of routes. These have already been recognized by Lam (1945) and others. The following distribution types can be distinguished:

(a) West: 22 species are found only in Sulawesi and west of it; they do not occur in the Philippines, Lesser Sunda Islands or New Guinea, suggesting that they reached the island by directly crossing Makassar Strait.

(b) North: 32 species occur in Sulawesi and the Philippines, but not in the Sunda block, or New Guinea.

(c) South: 27 species occur in Sulawesi and the LSI and/or Java, but not in Borneo, the Philippines, or the Moluccas.

(d) East: 41 species occur in Sulawesi, and New Guinea and/or the Moluccas, but not in the LSI, the Philippines, or the Sunda block.

(e) South-west: 60 species occur in Sulawesi, LSI, Borneo, and other areas west of Wallace's line. These species may have reached Sulawesi over the southern route or by direct crossing of the Makassar Strait.

(f) North-east: 21 species occur in Sulawesi and the islands to the east, and also in the Philippines. They may have come by way of the northern or the eastern route.

The connection to the west is again weakest, which is surprising considering the proximity and richness of Borneo. Apparently it has been much easier for many plants to reach Sulawesi by a northern, southern or eastern route.

Best represented in the west type are trees, mostly of montane habitats. In the north type most species are herbs of dry habitats; in the south type a large proportion of the species are herbs of dry and disturbed habitats, while most species of the east type are lowland forest trees.

DISCUSSION

The results of our analyses clearly show the importance of Wallace's line as a boundary, at least between Borneo and Sulawesi. Direct exchange between the two islands seems always to have been difficult. Our figures suggest that most plants have entered Sulawesi by northern, southern and eastern routes.

Within Malesia the strongest affinities of the Sulawesi flora at species level appear to be with the Lesser Sunda Islands (and Java), the Moluccas, and the Philippines. This can probably be ascribed, at least in part, to the effect of climate, since all these areas have large parts with a seasonal climate.

At genus level the affinity of Sulawesi is strongest with New Guinea. This is very difficult to explain from the present distribution of land and sea, for New Guinea is much more remote from Sulawesi than Borneo, even taking into consideration the Moluccan islands between Sulawesi and New Guinea. This renders wavering support to the idea of a westward rafting of parts of New Guinea.

No demarcation line was found between the western and eastern parts of Sulawesi, neither should it be expected. If the timing of the collision between the western and eastern part is correct, i.e. 15 Ma ago, it is only to be expected that any differences in floristic composition have become obliterated.

ACKNOWLEDGEMENTS

This chapter is based on a doctoral report by two of my students R. de Koning and M. S. M. Sosef. A fuller study is in preparation by them. It will contain further details on methods and results.

9 LEPIDOPTERA PATTERNS INVOLVING SULAWESI: WHAT DO THEY INDICATE OF PAST GEOGRAPHY?

J. D. Holloway

The geological scene set in Chapter 2, together with other possible hypotheses of the geological history of Sulawesi, provide a basis for predictions of biogeographical pattern. This is examined through reference to the Lepidoptera.

A phenetic analysis of butterfly distributions indicates that Sulawesi has its strongest association with the Philippines, reflecting the geological connection from Sulawesi through Sangihe to Mindanao. It also indicates a much greater degree of geographical isolation from Borneo than exists at present or than is suggested by some of the geological scenarios.

Two moth families, Notodontidae and Limacodidae, are examined in detail, including some phylogenetic analyses. Association with the Philippines is evident in both families. The fauna for both is almost entirely Oriental in character, old Australasian connections being virtually absent. Several endemic speciations are identified and discussed; elements of these show some degree of local segregation, though not specifically to one or other of the Oriental and Australasian geological components. The fauna as a whole appears uniformly distributed over the island, as seen also in the flora (this volume, Chapter 8).

Thus it is suggested that the hypothesis of the formation of Sulawesi from convergent Oriental and Australasian components is placed under two constraints: (a) the Oriental component of Sulawesi has, until recently, been considerably more isolated from Borneo than it has from the Philippines; (b) the Australasian component was submerged until fusion with the Oriental component at 15 Ma, and was thus unable to contribute old Australasian faunistic elements to the Sulawesi biota.

INTRODUCTION

To Alfred Russel Wallace (1880), Sulawesi 'presents us with a most striking example of the interest that attaches to the study of the geographical distribution of animals. We can see that their present distribution upon the globe is the result of all the more recent changes the Earth's surface has undergone; and by a careful study of the phenomena we are sometimes able to deduce approximately what those past changes must have been in order to produce the distributions we find to exist.'

Croizat (1958) (see also Craw and Weston 1984) suggested that within Sulawesi there was a definite break between Oriental and Australian biotic elements, these remaining distinct from each other in terms of local distribution, and supporting his idea of the composite geological nature of the island: 'Celebesian life is answerable not to the "landbridges" by which the modern island is geographically one, but to the "islands" . . . that discrete in the distant past, still are biologically discrete to-day within the body of Celebes.'

Both Wallace and Croizat were aware of the dynamic nature of the geographical history of Sulawesi, a history that has subsequently been clarified to some extent by modern studies of plate tectonics and stratigraphy. Yet there are still two somewhat conflicting hypotheses of the tectonic history of Sulawesi. In this paper these will be compared in the light of data from Lepidoptera distributions. Hypotheses of constant geography or of a vicariant sundering of a continuous extent of land from south-east Asia to Australasia through erosion (van Steenis 1962) or earth expansion (Cracraft 1980) are not compatible with the geography of the plants and animals found throughout the region (Holloway 1970, 1982b, 1984) and so will not be considered here.

GEOLOGICAL HYPOTHESES AND THEIR BIOGEOGRAPHICAL IMPLICATIONS

There has been disagreement in the geological literature over the composite nature of Sulawesi and the history of the component parts in the second half of the Tertiary in relation to each other and to lands to east and west. There is a consensus that the northern and southern limbs of the island are geologically distinct from the east and south-east limbs, the suture between the two halves of the island running approximately down the middle of the central portion between the Tomini and Bone bays.

Audley-Charles and colleagues (e.g. Audley-Charles 1974, 1981 and this volume, Chapter 2) have suggested that the western part has always been approximately in its present position relative to eastern Borneo and is an island arc of Sundaic origin linking to the north with Mindanao. The eastern part is of Australian origin, perhaps part of extensive formations associated with the Westralian geosyncline running through Sumba, Timor, Kai, Tanimbar, south-western New Guinea, and the southern Moluccas. The eastern portion, according to this hypothesis, has converged with the western part from a much more easterly position as a consequence of westerly thrusting through New Guinea along the Sorong Fault. This hypothesis would indicate that, though the two portions have converged to fuse from widely separated positions over the latter part of the Tertiary, their juxtaposition on the one hand to Borneo and the southern Philippines and on the other to the southern Moluccas and western New Guinea has remained comparatively constant.

Thus, if Croizat is correct and the biotas of independent geographical entities remain largely discrete when they converge to fuse and, if both sections were dry land from the earlier period, one might expect still to be able to trace a sequence of patterns of faunal interchange and speciation, the oldest of which would fall into discrete easterly and westerly components with Australasian and Oriental affinity respectively, but with more recent ones showing increasing degrees of exchange between east and west. If one or other component had a much shorter subaerial history than the other, then ancient east or west patterns would be absent, never having existed.

Even if lowland faunas of the two components have intermingled, patterns may persist in montane biotas though, as with those of oceanic islands, they may tend to have a predominance of dispersive elements where the mountains are geologically young (Smith 1977). Indeed, many montane moth species have a distribution pattern from Borneo through Sulawesi and the south Moluccas to New Guinea that reflects the present day 'central archipelago' of montane habitats in the Indo–Australian tropics (Holloway 1970, in press).

Another, now unfashionable, hypothesis (Katili 1978), whilst not discounting that some more easterly islands of the Sulawesi group such as the Sula islands and Buton (Pigram and Panggabean 1983) may be of Australasian origin, suggests that the two main components of Sulawesi represent a now-fused double arc system that developed south of the Philippines and well east of Borneo, both arcs being subsequently thrust west by the left-lateral shear through New Guinea. Again, the extent of land and its altitude during this period, and the degree to which the geological connection with the Philippines was manifest as dry land, is unclear.

On this second hypothesis older biogeographical patterns would reflect the isolation by a high degree of endemism and would exclude Borneo to a great extent, and perhaps New Guinea to a lesser extent. Borneo would feature increasingly strongly in more recent patterns of exchange and speciation as the two areas converged. Exchange with the Philippines would depend on the extent to which the Sangihe and Talaud structures were exposed, but might well precede extensive exchange with Borneo.

In the rest of this paper these alternative propositions will be examined in relation to general butterfly faunistics and to phyletic relationships within two moth families the taxonomy of which has been or is being studied in detail, namely the Notodontidae (Holloway 1983) and the Limacodidae Holloway 1986.

LEPIDOPTERA GEOGRAPHY

In a numerical classification of major islands in the Indo–Australian tropics in terms of similarity (shared species in relation to unshared) of butterfly faunas (Holloway and Jardine 1968), Sulawesi is associated primarily, albeit at high level, with the Philippines before association with Sundaland (Fig. 9.1). At a shared-species level Wallace's line (west of Sulawesi) is not the major discontinuity between the Australasian and Oriental butterfly faunas; this is the more easterly Weber's Line of Faunal Balance which runs between Sulawesi and the Moluccas. There is no significant discontinuity between Bali and Lombok, but the Wallace discontinuity separates the Philippines and Sulawesi from the rest of the Oriental region (Fig. 9.2).

A two-dimensional ordination (non-metric multidimensional scaling) gives an array of points (Fig. 9.1) representing the areas that has a general correlation with their geographical relationships. Therefore there is justification for interpreting departures from such correlation in terms of past geography (Holloway and Jardine 1968). In this analysis, and another one based on bird species data which gave very similar results, Sulawesi is distinctly displaced from the cluster of points that contains Borneo and the rest of the Sundaic land areas and appears to relate to these areas through the Philippines. This is concordant with the hypothesis of an originally isolated Sulawesi archipelago that has been thrust westwards towards Borneo through time. However, a similar result would be expected if there had been a collision (fide Audley-Charles): the consequent mingling of Oriental and Australasian faunistic components would cause Sulawesi to be placed in an 'average' position in the ordination, relatively isolated from neighbouring islands with purely Australasian or Oriental faunas. If, prior to collision, faunistic distances betwen the Oriental (O) and Australasian (A) parts of Sulawesi and the Moluccas (M) and Borneo (B) were OM and OB, and AM and AB respectively, then the post-collision distances between a united Sulawesi and its neighbours would be expected to be $\frac{1}{2}$(OM + AM) and $\frac{1}{2}$(OB + AB), OM and AB being of the order of magnitude of distance MB today.

Attention must therefore be directed to the Sulawesi fauna itself and the phyletic relationships of the species to those of neighbouring areas. This should reveal the extent to which there exist together faunistic components that could be considered purely Oriental or Australasian, particularly ones with indications of antiquity. Absence of an Australasian faunistic component would lend support either to the isolated archipelago hypothesis or to the possibility that the Oriental part of Sulawesi was at first in an isolated position relative to Borneo, with the Australian part being submerged until about the time of the collision.

One problem with such a phyletic analysis is that it is not easy to sequence the development of various distribution patterns in time. An *ad hoc* approach is to assume that complexity of pattern is correlated with geological age, the complexity being indicated by both the branching structure of the phyletic diagram and the degree of sympatry of the taxa (Holloway 1982*b*, 1984). Morphological divergence of endemics might be equated with period of isolation.

In previous papers (Holloway 1973, 1982*b*, 1984) I have drawn attention to allopatric arrays of taxa that extend through the Indo–Australian tropics, the component species of which tend to fall into a series of common distributional categories: mainland Asian, Sundaic, Wallacean, Lesser Sunda, Melanesian, and south-west Pacific (Fig. 9.2). The boundaries of these ele-

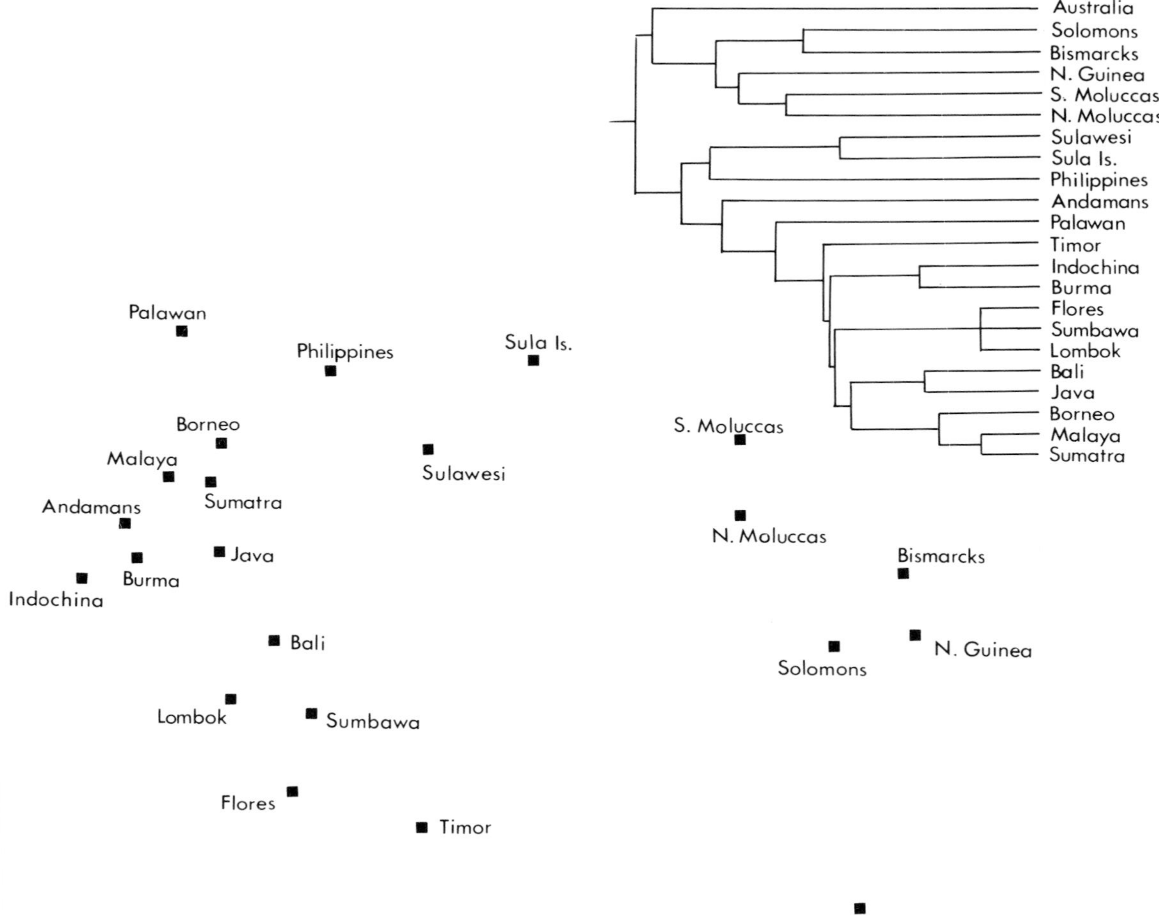

Fig. 9.1. Dendrogram showing relationships of areas in the Indo–Australian tropics in terms of a measure of the similarity of their butterfly faunas at species level, and a two-dimensional plot of points representing the areas derived through non-metric multidimensional scaling of the same data (after Holloway and Jardine 1968).

ments coincide with the major discontinuities in the faunistic classification of the areas discussed earlier. In the two moth families discussed here there are four examples, three from the Notodontidae and one from the Limacodidae (Fig. 9.3). This pattern is relatively simple and is taken therefore to be of comparatively recent origin, arising through the vicariant fragmentation of the range of an ancestor species that was distributed throughout the Indo–Australian tropics. In some instances the more westerly taxa overlap those to the east, presumably through a dispersal subsequent to the fragmentation, extreme examples of which may be found in the noctuid genera *Aplotelia* and *Paectes* (Holloway 1985).

Numerous species are today widespread through the Indo–Australian tropics, particularly in mobile families such as Noctuidae, and such species make up one-third to well over a half of the faunas of Pacific archipelagos from New Caledonia and Vanuatu eastwards (Holloway

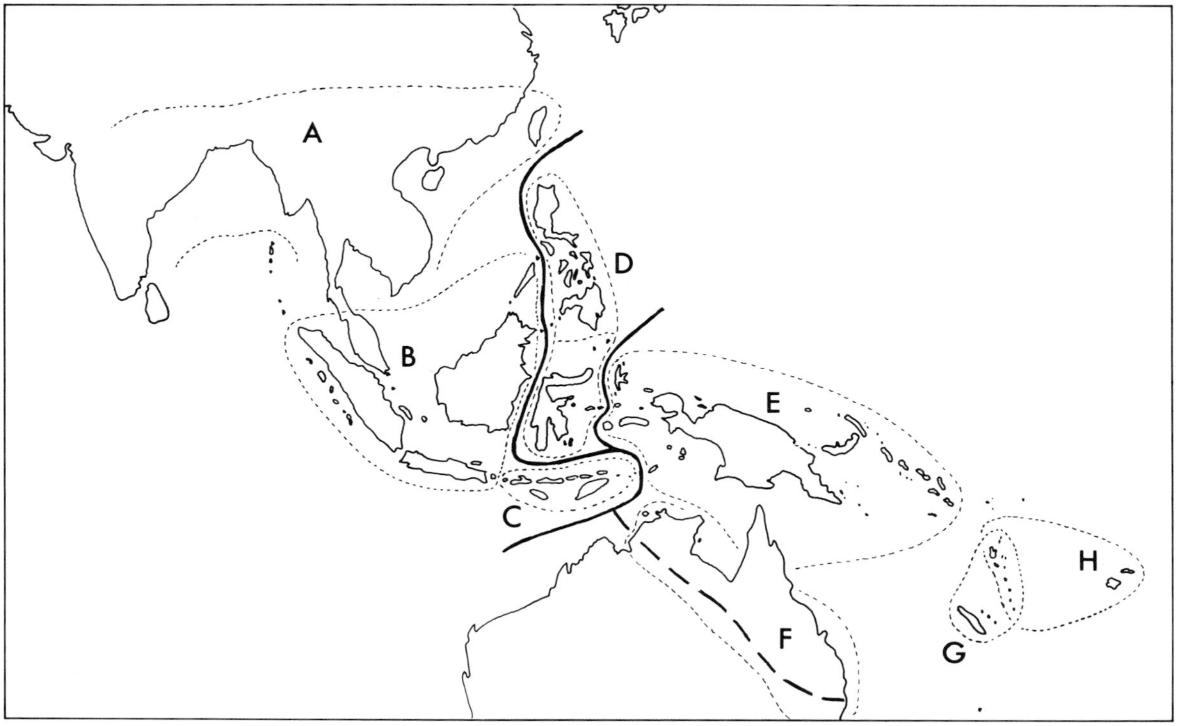

Fig. 9.2. Map of the Indo–Australian tropics showing (heavy lines, solid and dashed) major discontinuities in butterfly faunas at the species level and (fine dashed lines) frequently encountered species distribution patterns (Holloway 1973, 1979, 1982*b*) apart from endemics and extremely widespread species: A, mainland Asian; B, Sundaic; C, Lesser Sunda; D, Wallacean; E, Melanesian; F, tropical Australian; G, New Caledonian; H, Fijian.

Fig. 9.3. Generalized area cladogram (left) for allopatric arrays of species extending through the Indo–Australian tropics (from Fig. 9.2), and four examples of species groups from the two moth families Notodontidae and Limacodidae that follow this pattern closely.

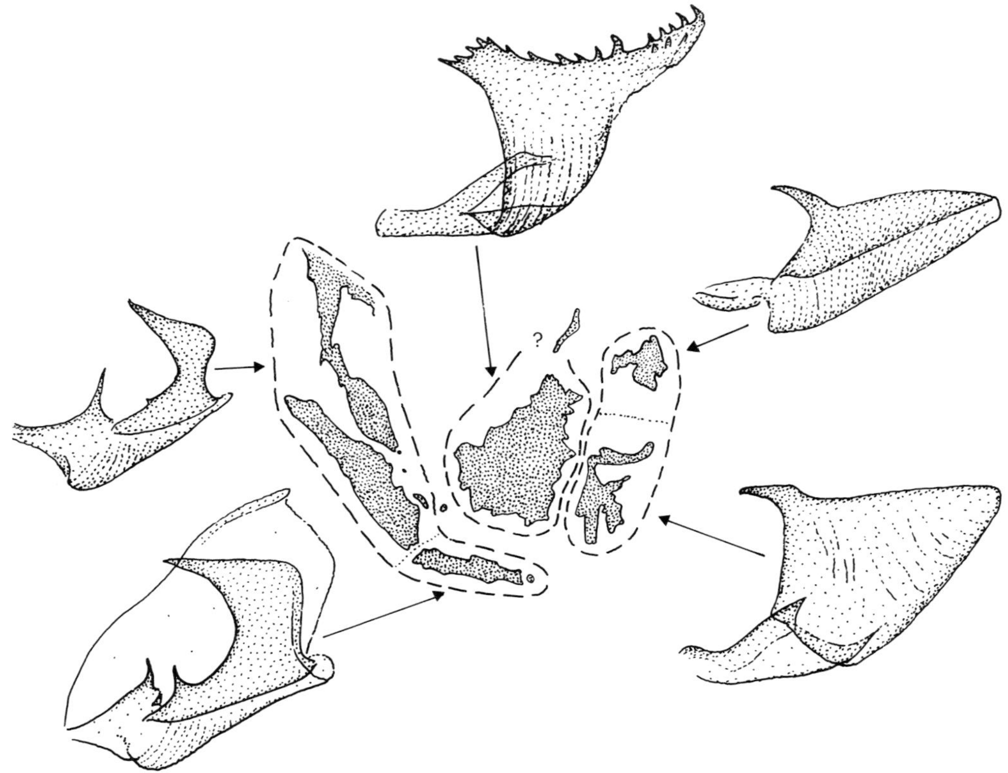

Fig. 9.4. Distribution of the moth *Besida xylinata* and its subspecies, showing characteristics of the valve ornamentation throughout the range. The complete valve is shown for the typical (Javan) race at lower left; the other diagrams show the saccular process only.

1979). Families or subfamilies with a high proportion of such taxa and with low island endemism are thus unlikely to be informative about most complex distribution patterns of more ancient establishment, as these patterns will have been blurred by subsequent dispersal.

Some generally less mobile families show patterns with resolution of areas within the major blocks involved in the allopatric arrays just described. They have overall a higher degree of endemism to individual areas.

The upper cladogram in Fig. 9.5 illustrates a pattern of area relationships that, to a greater or lesser degree of resolution, is common to at least seven groups of limacodid species (Holloway 1986). It is predicted to be the most likely pattern of breakdown of Sundaic species (see Fig. 9.3) into subspecies. It differs from the phenogram for the same four areas derived by Holloway and Jardine (1968) in that Borneo, rather than Java, is the outlier. The association of Borneo with Malaya + Sumatra in the phenogram may be due to the subsequent dispersal of the Malaya + Sumatra taxon to Borneo, increasing the overall similarity between the three areas.

Given the assumption that these area relationships represent a single sequence of vicariant fragmentation of a widespread Sundaic taxon, it is possible to use it to sequence relationships involving Sulawesi. In both the two other cladograms (Notodontidae) in Fig. 9.5, for subspecies of *Besida xylinata* (Fig. 9.4) and allied genera, and for subgenus *Erconholda* of *Phalera* (Holloway 1983), there is an exclusive relation-

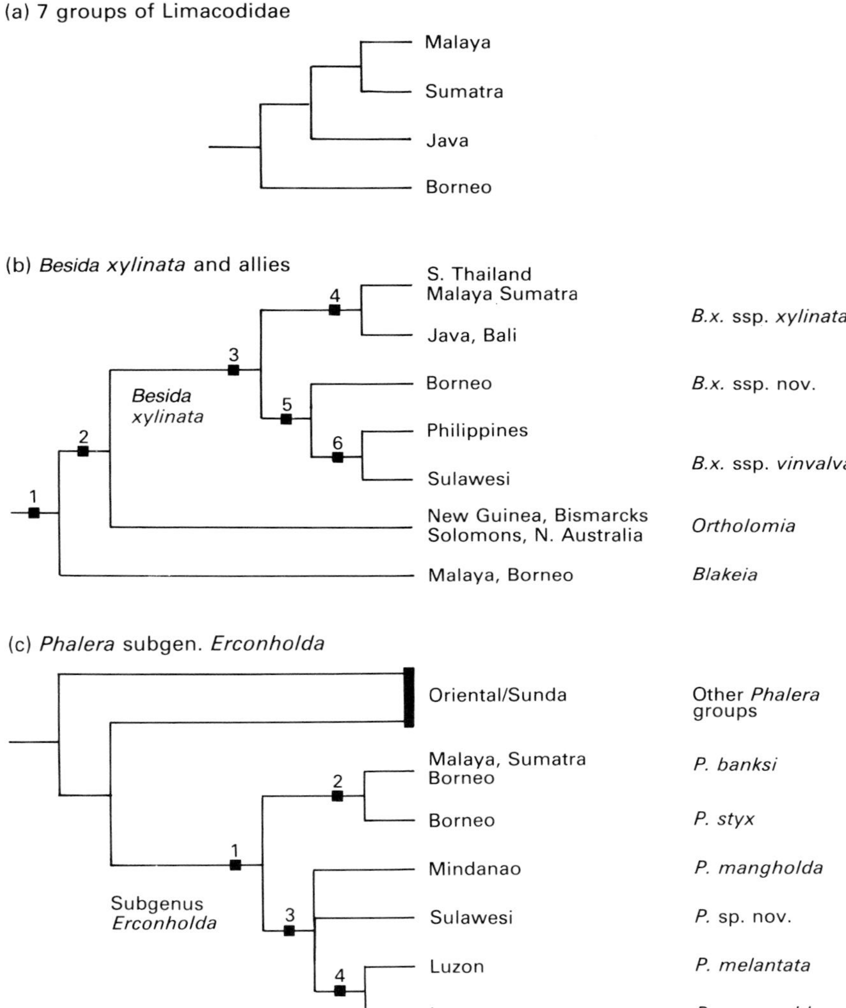

Fig. 9.5. (a) general area cladogram for the major land areas of Sundaland, derived from seven groups of Limacodidae; (b) cladogram or *Besida xylinata* and allied genera, the groupings defined by presumed apomorphies as listed in Appendix 4; (c) cladogram for subgenus *Erconholda* of *Phalera*, the groupings defined by presumed apomorphies as listed in Appendix 4.

ship between Sulawesi and the Philippines. In the *Besida* example (Fig. 9.5b) it occurs within the Sundaic cladogram as sister to Borneo, whereas in the *Erconholda* example (Fig. 9.5c) it is sister to the whole Sundaic cladogram (though this excludes Java and includes sympatry in Borneo). The *Erconholda* cladogram is compatible with the area relationships shown between the areas concerned in the allopatric Indo–Australian groups of Fig. 9.2. The *Besida* example suggests that zoogeographical associations between Philippines and Sulawesi, as strong as or stronger than that between either place and Borneo, persisted beyond the development of the Sundaic area pattern. In neither example is it possible to determine whether, if the original wide range of

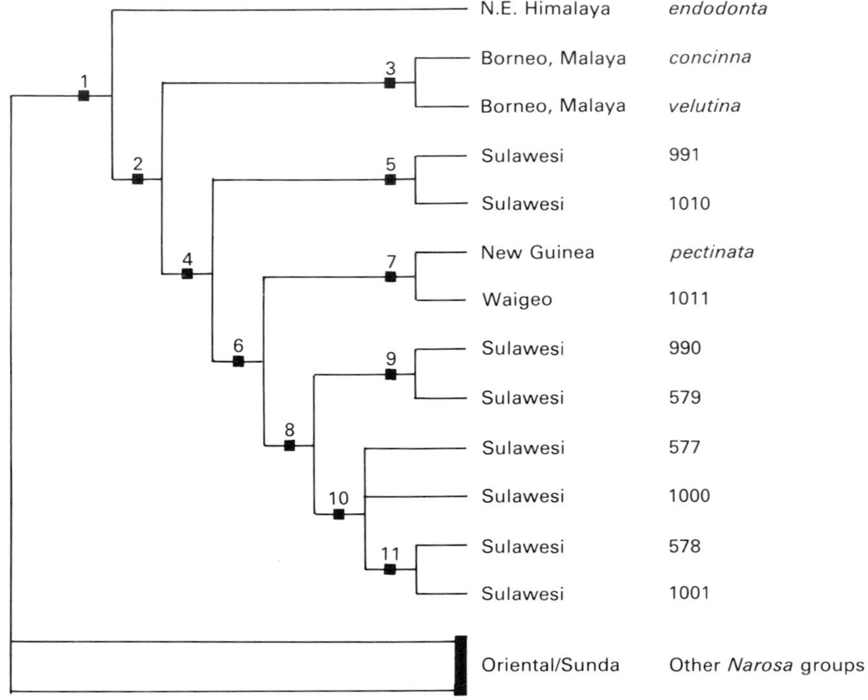

Fig. 9.6. Cladogram for the *velutina* group of the limacodid genus *Narosa*, the groupings defined by presumed apomorphies as listed in Appendix 4. The numbers at some branch ends refer to BMNH genitalia slide numbers for undescribed species. One further species has been added to each of the Sulawesi groups defined by apomorphies 5 and 9 (Project Wallace material from north Sulawesi), but these were discovered too late for inclusion.

the ancestral taxon arose through dispersal from Borneo, this was via the Philippines or Sulawesi. But the general area relationships shown by the Holloway and Jardine (1968) analysis suggest that Borneo to Philippines is the most likely route.

Both the cladograms (Figs. 9.5b, c) are from the family Notodontidae. The Notodontidae and Limacodidae show a relatively high degree of endemism to areas within the Indo–Australian tropics and are represented in moderate diversity within both the Oriental region and Australasia.

It would, of course, be most apposite to follow on from the area analysis on butterflies discussed above and shown in Figs 9.1 and 9.2 by a discussion of phyletic patterns within that group. Unfortunately to do so is premature because very few phyletic generic analyses within the butterflies have yet been made. The collections made by Project Wallace during 1985 have yet to be analysed. Similar patterns to those found in the two moth families can be anticipated: (1) Philippines–Sulawesi links are evident in the genera *Moduza*, *Acropthalmia*, and *Zethera*; (2) monospecific isolated endemic genera are numerous e.g. *Aoa*, *Nirvana*, *Pseudomycalesis*, and *Bletogona*; (3) an endemic radiation is seen in *Lohora* (perhaps only a section of *Mycalesis*); (4) a complex sister relationship between an Oriental (*Faunis* and allies) and an Australasian (*Taenaris* and allies) genus is seen virtually to exclude Sulawesi (one *Faunis* species only is recorded, possibly allied to Philippines species); (5) Australasian relationships are seen in

Elodina. But the butterfly fauna as a whole is overwhelmingly Oriental in character (Holloway 1973).

We now turn to a detailed examination of the distribution patterns of the Sulawesi species of the two particular moth families. A selection of species is shown in Fig. 9.7.

The moth family Notodontidae

The family Notodontidae has at present 52 species recorded from Sulawesi, listed with some taxonomic notes in Appendix 1. The fauna is much less diverse than that of Sumatra (154 species), Borneo (124 species), or New Guinea (100 species). About 40 per cent are endemic but most of these have close relatives in neighbouring islands or islands groups.

The overall picture is as follows, the letter-codes referring to Appendix 1 which gives full details:

Thirteen species (**a**) are found widely in the Oriental tropics from India eastwards. Of these, six (**aa**) extend into the Australasian tropics and two more (**aaa**) have close relatives there.

Five species (**b**) are found also in most islands on the Sunda shelf; half are in the Philippines. In most instances the Sulawesi population is subspecifically distinct, but in the case of *Besida xylinata* (discussed earlier; Figs 9.4, 9.5, 9.7) the Sulawesi subspecies occurs also in the Philippines.

Eleven of the Sulawesi endemic species (**c**) have sister or close relationships with Sundaic species, four of which are also in the Philippines.

One endemic species (**d**) has allies in the Philippines, Sumatra and Bali, but the genus is absent from Borneo.

Five species (**e**) belong to more or less widespread allopatric arrays of species of the type discussed earlier (Fig. 9.3). In two instances the Sulawesi species is shared exclusively with the Philippines.

In addition to three species listed under (b) and (e), another eight (**f**) occur both in Sulawesi and the Philippines, or have their closest relatives in the latter; all are of Oriental affinity except one (**ff**).

Five species (**g**) have mainly eastern associations, although four of these have more distant affinity westwards.

Finally, there are five endemic species (**h**), all distinctive and all with no obvious sister relationships.

In addition to these affinities, limited endemic radiation is evident in the Sulawesi Notodontidae: *Loda celebensis* (Fig. 9.7) and *Melagona dentatus* (Fig. 9.7) are monobasic endemic genera that belong to the *Chadisra* group (Holloway 1983) and are here subordinated as subgenera to *Chadisra* in Appendix 2 (which sets out a tentative rearrangement of taxa in the *Chadisra* group that occur in the Indo–Australian tropics). The distribution of the group as a whole has the appearance of a relatively old dispersal through the area with divergence in isolation, followed by further episodes of dispersal, overlap and speciation. The two endemic subgenera in Sulawesi share an interior orientation of the apophyses of the ovipositor lobes in association with a deep intersegmental membrane and thus could be sister-taxa. Outside this the relationships are obscure because of the extreme and diverse modifications of male and female genitalia structure exhibited by the group.

The sister-pair of *Higena* species provide a clear instance of endemic speciation (Fig. 9.7) which has possibly also occurred in the genus *Ambadra* (Fig. 9.7). Species of this latter genus are of very uniform morphology and so phyletic relationships are unclear, but possibly with the Philippines fauna. The species are palm-feeders and so the abundance of palms, including rattans in the natural forests could well have provided a stimulus for speciation.

In summary, the Notodontidae show only limited endemic speciation and strong western affinities. Just over 20 per cent of the fauna has a primary relationship exclusively with the Philippines, though this includes the *Ambadra* species.

The moth family Limacodidae

This has at present 33 species recorded from Sulawesi, compared with 95 from Borneo. The

Fig. 9.7. Various of the Sulawesi endemic moths discussed in the text (top to bottom). Family Notodontidae (left column: *Ambadra celebensis*, *A. ?suriga*, *A. 1659*, *A. 806*, *Higena 1425*♂, *H. 1425*♀, *H. 1721*; right column: *Chadisra (Loda) celebensis* 2 forms, *C. (Melagona) dentatus*, *Lasioceros euteles*, *Besida xylinata*, *Phalera*, subgen. *Erconholda* sp. nov.). Family Limacodidae (centre column) major species radiation of *Narosa* (spp. *1044*, *1010*, *991*, *577*, *579*, *578*, *1001*, *990*, *1045*) (see also Fig. 9.6).

taxonomy of the family is under revision, with many generic changes imminent.

The overall picture is as follows, the letter-codes referring to Appendix 3 which gives fuller details.

None of the Sulawesi species is widespread in the Oriental tropics. One (**a**) extends from Sundaland to New Guinea, and two (**b**, **c**) are shared with Sundaland and, one only, with the Philippines.

There is one widespread allopatric array (**d**) in the genus *Birthamoides* (discussed earlier, Fig. 9.3); the Sulawesi species is shared with the Philippines.

Five species (**e**) have very close relatives in the Philippines, with which two more (**ee**) are shared.

There are seven further endemic species (**f**) of which five (including possibly three new genera) await description. Of these, five (**ff**) have Sundaic/Philippines affinities and the remaining two are morphologically isolated.

Three species (**g**) are primarily of eastern affinity, but belong to genera most diverse in the Oriental region.

The Limacodidae of Sulawesi include one example (**h**) of a major endemic radiation that also involves New Guinea. This is within the otherwise largely Oriental genus *Narosa*, see Fig. 9.7. Most of the species are undescribed and a first assessment of their relationships is presented in Fig. 9.6. The Sundaic pair (3), Sulawesi pair (5), and Sulawesi/New Guinea group (6) are all defined by strong apomorphies though their relationships with each other are not so clear. Alternatives to the relationships shown ((5,6) 3) are: ((3,5) 6), defined by elongation and sclerotization of the valve saccular processes; ((3,6) 5), defined by presence of a sclerotized manica plate. All possibilities involve homoplasy or reversal in the other characters.

The *Narosa* radiation appears to exclude the Philippines but species may await discovery there and also in the south Moluccas. The diverse cicada group reviewed by Duffels (1983), with a complex in Sulawesi sister to Inner and Outer Melanesian Arc groups, also excludes the Philippines, though the ancestor may have originated from there. The Sulawesi cicada group shows some local allopatry and this may also occur in *Narosa*. The *Narosa* group is only weakly represented in western New Guinea.

A minor radiation of three species is seen in the genus *Parasa*. One species is widespread, one only known from the north and the third only from the south-west. The trio forms a group with a Philippines species and a Sundaic one, both rather distinctive. The quintet in turn is related to the Oriental *lepida* group.

In summary, Sulawesi Limacodidae are probably of entirely Oriental affinity. Five genera extend further eastwards; otherwise the Australasian tropical limacodid fauna appears not to be closely related to that of the Oriental region, but has yet to be studied in detail. About 25 per cent of the fauna has a primary relationship exclusively with the Philippines.

DISCUSSION

The Limacodidae of Sulawesi include far fewer widespread Oriental species than do the Notodontidae, and a much larger proportion of endemics, about 66 per cent compared with 40 per cent. This may be taken as an indication of lower dispersive powers for the Limacodidae as a whole compared with the Notodontidae. Also in the Limacodidae Sundaic associations are fewer and Philippines ones relatively more numerous. Endemic speciation has been more extreme. A recent study has shown that in another moth family, the Noctuidae, the Sulawesi fauna is even more widespread Oriental than are the Notodontidae (Holloway 1985).

In the development of an island fauna both area and distance of potential source areas are of importance in determining faunistic relationships, but the effect of area decreases relative to distance at lower powers of dispersal. These predictions arise from the MacArthur and Wilson (1967) exponential model of dispersal which was tested and supported by data from the moths of Norfolk Island (Holloway 1977). The proportion of recent versus ancient patterns of distribution and affinities also increases with overall dispersal powers within a group.

The modern geographical proximity of Sulawesi to Borneo would lead one to expect a faunistic distance between them similar to those between the islands from the Moluccas to the Solomons where similar ocean gaps are involved (cf. Holloway and Jardine 1968). In fact the faunistic distance is much greater and the degree of endemism very much higher. Both these facts point to a greater water gap between the two at some time in the past rather than the constant juxtaposition indicated by Audley-Charles.

The old endemic elements must have established during such a period of isolation, often with no counterpart in the Philippines fauna as in *Chadisra*, *Narosa*, and some cicadas. The next association appears to have been predominantly with the Philippines and has continued to the present. Association with the Sundaic fauna appears to be the most recent and is much weaker in groups of low dispersive powers. The affinities of the oldest endemics are obscure, albeit broadly Oriental. Nor can it be clear whether they owe their original isolation to dispersal across a major water gap or to a vicariance event such as opening of the Makassar Strait between Sulawesi and Borneo.

Association with lands to the east appears to have been relatively weak throughout the history of Sulawesi where old and complex patterns span the Indo–Australian tropics, such as in the cicadas studied by Duffels (1983) and in the *Chadisra* group (Appendix 2). In both, the Sulawesi representatives offer evidence of development in independence from either east or west (except in the latter for some close and therefore probably recent sister-relationships).

Finally, what evidence is there for local disjunction of distribution within Sulawesi as suggested by Croizat? Analysis is still in progress of Lepidoptera collected in the past few years in various parts of the island. First impressions are that disjunctions do not occur, except in species groups such as the cicada genera and the *Narosa* complex of moths that have evolved within the island. The data do not indicate segregation of Australasian and Oriental components; the *Narosa* group, shared with New Guinea, is found throughout the island although is perhaps most diverse in the west and south-west.

The evidence from Lepidoptera geography therefore requires a much greater water gap between Sulawesi and Sundaland, particularly Borneo, than is manifest in the geological reconstructions presented by Audley-Charles (his Fig. 2.8, for example). It also indicates, as does the floristic evidence (van Balgooy, this volume, Chapter 8), the biological unity of Sulawesi, suggestive of a subaerial history as a single island or close-knit archipelago as suggested by Katili (1978). Only endemic species groups show segregation within the island but they occur throughout.

The hypothesis of Katili is not borne out by the geological evidence as reviewed by Audley-Charles, but perhaps the biological data allow certain constraints to be placed on a hypothesis of geological heterogeneity in Sulawesi. The requirement of greater isolation has already been mentioned. The absence of an old Australasian component and the biological unity of the island suggests that the Australasian part commenced its history as dry land more or less at the time of fusion of the two parts in the middle Miocene. This supports the geological evidence described by Audley-Charles. If the western part has had a longer subaerial history, then this must have involved a period of greater isolation from Sundaland than at present. But the biological evidence cannot indicate whether this greater water gap has been closed in time through convergence or through marine regression in eastern Borneo.

It is likely that the current interest in Sulawesi entomology through Project Wallace (1985) and Operation Drake (1979) will lead to detailed taxonomic research on a range of insect groups. This will provide more data, more biogeographical patterns to compare with the geological hypotheses of the time.

ACKNOWLEDGEMENTS

This paper (Results of Project Wallace No. 1) is based in part on material collected whilst the author was a participant on Project Wallace, sponsored by the Royal Entomological Society of London and the Indonesian Institute of Sciences (LIPI). Thanks are due to the British Ecological Society for a Travelling Fellowship which covered a large part of the author's expenses. Fig. 9.7 was made by Alan Wood.

Appendix 1. Preliminary checklist of Sulawesi Notodontidae (51 species) with annotations. Reference is given to the most recent literature on the species or species group. Endemics are asterisked.

aa *Tarsolepis sommeri* Hübner (Holloway and Bender, 1985).

e *Dudusa celebensis* Roepke* (Holloway 1983, p.19).

c *Gargetta* sp. near *hampsoni* Schintlmeister* (Holloway 1983, p.20).

f *Gargetta tompoa* Kiriakoff (Holloway 1983, p.21).

ff *Lasioceros euteles* West (Holloway 1979, p.362).

c *Porsica sidaonta* Kiriakoff*. Possibly allied to the widespread Oriental *P. punctifascia* Hampson.

a *Porsica curvaria* Hampson (Holloway 1983, p.26).

a *Phycidopsis albovittata* Hampson (Holloway 1983, p.27); also in Mindanao.

b *Besida xylinata* Walker ssp. *vinvalva* Schaus (see Figs 9.4, 9.5).

aaa *Archigargetta viridigrisea* Hampson (Holloway 1983, p.29).

a *Gangarides rosea* Walker (Holloway 1983, p.31).

e *Phalera sulawesiana* Holloway* (1982*a*, p.201, 1982*b*, Fig. 3).

f *Phalera* sp. n.* of subgenus *Erconholda* Kiriakoff (Holloway 1983, p.37 and this volume, Fig. 9.5).

b *Neodrymonia apiculatus* Rothschild ssp. *celebensis* Roepke **stat. n.** The generic combination was established by Holloway and Bender (1985). The Sulawesi race differs from typical *apiculatus* (recently recorded (C. G. Treadaway Coll.) also from the Philippines) in characters of the male genitalia: the broader costal process to the valve; the absence of a spur at the apex of the costal margin; the longer, more tapering uncus and the narrower, triangular fused gnathi.

f *Ambadra celebensis* Roepke **sp. rev.*** (Holloway 1983, p.40). This taxon was placed as a synonym of the Philippines *suriga* Schaus but the next species appears to be closer to *suriga*.

f *Ambadra* sp. ? *suriga* Schaus. No Philippines material is available to check this association.

f *Ambadra* sp. n.* (slide 806). A species smaller than the previous with white rather than pale reddish-brown hindwings.

f *Ambadra* sp. n.* (slide 1659). A species as large as *celebensis* but with distinctive facies and male genitalia.

c *Norraca celebica* Kiriakoff*. This species is allied to the Sundanian taxa *sordida* Roepke and *sabulosa* Kiriakoff, and possibly to *ubalvia* Schaus from the Philippines.

f *Pantanopsis celebensis* Gaede* (Holloway 1983, p.49). Also in Mindanao.

e *Cerura minahassae* Holloway (1982*a* p. 203, 1982*b*). This species has recently been recorded from the Philippines (C. G. Treadaway Coll.).

c *Neocerura* sp.*. This is possibly a subspecies of *liturata* Walker and certainly in a group with that widespread Oriental species and the Philippines species *hapale* West (Holloway 1982*a*, p.204).

g *Quadricalcarifera dasychirinus* Roepke* (= *paranga* Kiriakoff, **syn. n.**). This is a very variable montane species that belongs to a group that includes *grisescens* Roepke (= *ardjuna* Kiriakoff) from Java and Bali and has its centre of diversity in New Guinea and the S. Moluccas with at least the following species: *melanogramma* Joicey and Talbot (probably = *kebeae* Bethune-Baker, *mediobrunnea* Bethune-Baker, *dubiosus* Bethune-Baker, *rufotegula* Gaede); *aeruginosus* Gaede; *frugilegus* Rothschild; *ceramensis* Kiriakoff; *viridigriseus* Rothschild. Most of these taxa have a band of small spines lining the rim of the aedeagus. The group is related to that of the next species listed and to a large group of smaller, more delicate Australasian species that includes *choriolus* Joicey and Talbot, *nitidus* Rothschild, *purpurascens* Rothschild, *flavicollis* Rothschild, *rufescens* Rothschild, *dinawa* Bethune-Baker *medio-griseus* Gaede and *roseus* Gaede. It is likely that all Australasian *Quadricalcarifera* can be assigned to one or other of these two groups.

c *Quadricalcarifera celebensis* Roepke* (= *ferrea* Kiriakoff and *rhypara* Kiriakoff, **syns. n.**). It is likely that in the description of *ferrea* Kiriakoff (1967) transposed the illustrations of the male genitalia of *ferrea* and *triguttata*. A photograph in the BMNH of the holotype has been matched with a specimen with genitalia as in his Fig. 20. The variable species is common in the lowlands. The structure of the male genitalia, particularly of the aedeagus with a subapical spur, indicates a sister-relationship with the Philippine and Sundanian *charistera* West.

c *Quadricalcarifera alboviridis* Kiriakoff*. This is the sister-species or merely a subspecies of the Philippine and Sundanian *Q. palladina* Schaus.

h *Quadricalcarifera* sp.* (slide 1248). This species has distinctive male genitalia and cannot be associated closely with any congener examined.

aa *Vaneeckeia pallidifascia* Hampson (Holloway 1983, p.57).

c *Vaneeckeia* sp. n.

aa *Neostauropus alternus* Walker (Holloway 1983, p.58).

g *Neostauropus* sp. ? *amboynica* Oberthur. The Moluccan holotype of *amboynica* is female and only males have been taken in Sulawesi.

aa *Netria viridescens* Walker (Holloway 1983, p.61).

a *Somera viridifusca* Walker (Holloway 1983, p.62).

a *Formofentonia orbifer* Hampson (Holloway 1983, p. 63).

h *Higena* Matsumura (= *Sagamora* Kiriakoff, **syn. n.**, and *Kikuchiana* Matsumura) sp. n.* (slide 1425). In facies the male resembles that of the Sundanian *Sagamora indigofera* Holloway (possibly closely related to *H. biarcuata* Gaede (India) and *H. plumigera* Matsumura (Taiwan)) and the female that of *biarcuata*, but the male genitalia are very different.

h *Higena* sp. n.* (slide 1721). A blackish species which, on similarity of male genitalia, is probably sister to the previous one.

g *Omichlis* sp. n.* (slide 1426). The genitalia resemble those of some New Guinea species of this mainly Australasian genus.

b *Calyptronotum singapura* Gaede ssp. n. The male genitalia of Sulawesi specimens differ in uncus structure from typical Sundanian *singapura* and the Philippines subspecies *gualberta* Schaus **stat. n.**

d *Maguila* Kiriakoff (= *Chloroceramis* Kiriakoff, **syn. n.**) ? sp. n.* (slide 1471). Sulawesi material does not differ markedly from *M. maguila* Schaus (Philippines) or *M. viridinota* Hampson **comb. n.** (N.E. Himalaya, Sumatra, Bali).

g *Oxoia irrorativiridis* Bethune-Baker (= *triangularis* Gaede, **syn. n.**) (see also Holloway 1983, p.67). The male genitalia of the two taxa synonymized are almost identical.

aa *Neopheosia fasciata* Moore (Holloway 1983, p.69).

c *Hyperaeschrella* sp. n.* (slide 1472). This is allied to two Sundanian species discussed by Holloway (1983, pp.72–3).

b *Pseudohoplitis vernalis* Gaede (Holloway 1983, p.77).

g *Chadisra (Sawia) vittata* Kiriakoff*. This species is allied most closely to *undulata* Kiriakoff (New Guinea), then to the other taxa in Appendix 2 listed for the subgenus.

c *Chadisra (Chadisrella) celebensis* Kiriakoff* (Holloway 1983, p.80).

h *Chadisra (Loda) celebensis* Kiriakoff* **comb. n.** See Appendix 2. This combination brings about a secondary homonymy such that a new name is needed for the previous species. This must await further detailed study of the *Chadisra* complex.

h *Chadisra (Melagona) dentatus* Gaede* **comb. n.** (= *simplificata* Gaede, **syn. n.**). See Appendix 2.

e *Teleclita cathana* Schaus (Holloway 1983, p.81 and this volume, see Fig. 9.3).

b *Allata argentifera* Walker (Holloway 1983, pp.85–6).

aa *Allata* sp. ? *indistincta* Rothschild (Holloway 1983, p.86).

aaa *Clostera restitura* Walker (Holloway 1983, p.90).

c *Clostera* sp. n.* in *dorsalis* Walker group (Holloway 1983, pp.92–3).

e *Micromelalopha celebesa* Tams* (Holloway 1983, p. 95).

Appendix 2. Checklist of subgenera of the *Chadisra* Walker generic complex of moths (see also Holloway 1983, p.78).

Chadisra Walker	
bipars Walker	N.E. Himalaya, Sumatra
borneensis Holloway	Borneo
sp. n. (slide 1949)	Java
Subgenus n.	
sp. n. (slide 1106)	S. India
Trincomala Kiriakoff	
basalis Moore	Sri Lanka, S. India
Stenoschachia Matsumura	
bipartita Matsumura	Taiwan, Sumatra, Java
Subgenus n.	
kiriakoffi Holloway and Bender	Sumatra, ? Peninsular Malaysia
Chadisrella Kiriakoff	
basivacua Walker	Sundaland, ? Philippines
celebensis Kiriakoff	Sulawesi
Sawia Kiriakoff	
luzonensis Kiriakoff	Sundaland, Philippines
sp. n. (slide 1107)	Philippines (Luzon)
vittata Kiriakoff	Sulawesi
undulata Kiriakoff	New Guinea
Loda Kiriakoff	
celebensis Kiriakoff	Sulawesi
Melagona Gaede	
dentatus Gaede	Sulawesi
Timoraca Kiriakoff	
meeki Rothschild	New Guinea
Antithemerastis Kiriakoff	
striata Rothschild **comb. n.**	S. Moluccas, New Guinea, Bismarcks, Queensland
acrobela Turner (? = *striata*)	Queensland
hendersoni Kiriakoff	Solomons
Parachadisra Gaede	
varians Bethune-Baker	New Guinea
[*atrifusa* Hampson	N.E. Himalaya]

Appendix 3. Preliminary checklist of Sulawesi Limacodidae. Further information on the status of some of these species will become available on the publication of monographs in preparation on the Bornean fauna and on genera and species complexes that are pests of coconuts. Endemics are asterisked.

ff *'Altha' alastor* Tams*. A taxonomically isolated species misplaced in *Altha*.

h *Narosa* Walker: 10* undescribed species in two groups as discussed in the text and shown in Fig. 9.6.

a *Chalcocelis albiguttatus* Snellen.

e *Parasa chlorostigma* Snellen* (= *vinculum* Hering).

e *Parasa*: 2* new species related to *chlorostigma*.

ee *Scopelodes magnifica* Hering. This is also known from the Philippines (C. G. Treadaway Coll.) and may be allied to the New Guinea *S. nitens* Bethune-Baker.

g *Scopelodes exigua* Hering. This is a S. Moluccan species allied to the Sundaic *unicolor* Walker and *dinawa* Bethune-Baker from New Guinea.

e, g *Thosea* Walker: 3* species, one probably allied to a pair from the Philippines and two to *monoloncha* Meyrick and allies from Australasia.

b *Thosea vetusta* Walker.

ee *Praesetora irrorata* West.

d *Birthamoides circinatus* Snellen (= *bilineata* West). See Fig. 9.3.

e *Setora* sp.n.* allied to two undescribed Philippines species, then to mainland Asian taxa, *not* to the Sundanian *nitens* Walker group.

c *Birthama congrua* Walker.

ff *'Thosea'*: 2* species allied to *'T.' cruda* Walker (Sundaland) and to be placed in a new genus.

e *'Thosea' porthetes* Tams*. To be placed in a new genus with allied Philippines species.

ff *'Miresa'* sp. n.* allied to *sola* Swinhoe, to be placed in a new genus.

ff *Darna catenatus* Snellen*. Distantly related to the *D. trima* Moore complex.

f 2* taxonomically isolated species of uncertain affinity.

Appendix 4. Presumed apomorphies used in construction of the cladograms in Figs 9.5 and 9.6.

Besida xylinata and allied genera

1. Forewing facies in common, especially a pale bar bisecting the tornal angle.
2. Broad valve in male genitalia with saccular and central processes.
3. Only saccular process of valve prominent.
4. Saccular process sickle-shaped (Fig. 4).
5. Saccular process triangular with interior angle somewhat spurred; aedeagus apex produced.
6. Interior spur of saccular process single, dorsal margin smooth.

Subgenus *Erconholda* of *Phalera*

1. Uncus of male genitalia massive, globular.
2. Forewings relatively narrow, with distal margins less oblique than usual in *Phalera*.
3. Forewings more triangular than in *Phalera* generally; ventral angles of uncus produced ventrad; gnathi (or socii) set obliquely on uncus face.
4. Valve with all three interior processes small, the most basal set distally relative to position in congeners.

P. mangholda and the Sulawesi species have spines near the aedeagus apex that may be homologous and indicate a sister-relationship.

Narosa velutina group

1. Valve of male genitalia expanded through upward torsion.
2. Valve bilobed.
3. Sclerotized manica plate present, bifid; saccular processes of valve distally scobinate, and well, but asymmetrically, developed (the saccular process in *endodonta* is simple, rod-like).
4. Male antennae bipectinate (a reversal; filiform antennae seen in most *Narosa* and allied genera are atypical in limacodids).
5. Wings whitish; hindwing with dorsum darkened; manica plate absent; saccular processes developed symmetrically into hooks.
6. Apical rim of aedeagus expanded, trumpet-like; manica plate quadrifid; saccular process of valve reduced, unsclerotized.
7. Broad, rather square saccular process.
8. Basal lobe of valve larger than dorsal one (vice versa in sister-group and out-groups).
9. Central two prongs of manica large, with connecting lamina extending further dorsal than that between the centre and exterior prongs.
10. Central manica prongs very small or absent.
11. Central manica prongs arising from base of exterior prongs.

10 COMPLEX ORIGINS

Wilma George

A SHORT HISTORY OF BIOGEOGRAPHY

Biogeography can be studied in two different ways. It can be studied by plotting the distribution of modern floras and faunas, classifying them, comparing them, and analysing their interrelationships. Or, it can be studied by looking at the history of the floras and faunas and the history of the land on which they live.

In the early nineteenth century, biogeographers, becoming aware of the immense diversity of floras and faunas that were being discovered all over the world, attempted to classify and compare. To do this they needed a system. One of the earliest systems divided the world into vegetational zones. Alexander von Humboldt (1769–1859), impressed by the altitudinal zones of plants on the slopes of Mount Chimborazo in Ecuador, inferred a similar zonation of plant communities through the latitudes of the earth: 'the relation, above alluded to, between the absolute height of the ground and the geographical as well as isothermal latitude, shows itself often' (Humboldt 1849). He compared the lichens and mosses of the mountain peak to the vegetation of the polar regions, the lower birch and shrub zone to northern tundra. The vegetational zones were distinct plant communities or, as he called them, plant nations. For Humboldt, altitude and latitude with their consequent climates played the major role in the determination of plant distribution. This view was transferred to spiders and insects when Pierre Latreille (1762–1833) wrote a geography of those animals.

'Botanical arithmetic' was another system of biogeography. Augustin-Pyramus de Candolle (1778–1841) counted the families and species of plants in different regions and counted the number of species in a genus. He arrived at the curious observation that the more species there were in a genus the more monotonous the flora of an area, while the fewer the species, the more diverse the flora. He compared the 7.2 species per genus of the French flora to the 1.5 of the Canaries and declared the Canaries the more interesting. He would have found the palms of the eastern tropics at 14.4 species per genus (Dransfield, Chapter 6) exceedingly monotonous but the non-volant mammals of Sulawesi with 1.7 species per genus (Musser, Chapter 7) excitingly diverse. Jean-Théodore Lacordaire (1801–70) compared the insect faunas of the world in the same way. The index of diversity for South American insects worked out at 617 species per genus, monotonous compared with New Holland (coastal Australia, Tasmania, and New Zealand) at 2 species per genus. His table of beetle distribution also showed that South America had ten times as many species as New Holland and three times as many as the Indian archipelago. However, as he realized, this was not necessarily a genuine comparison as New Holland species were not well known. In addition, numbers of beetle species were different in different parts of the world, but different families were dominant: lamellicorns (scarabeid chafers, lucanid and stag beetles) dominated the fauna of New Holland, curculionids (weevils) the Indian archipelago, and carabids (ground beetles) the Asian mainland (Fig. 10.1).

The significance of Humboldt's nations and de Candolle's indices of diversity was that they made it possible to compare qualitatively different areas of the world. These early biogeographers were floral and faunal pattern-makers. Biological arithmetic continues to be used for pattern-making, vegetational zones still classify and make tropical, temperate, and polar pat-

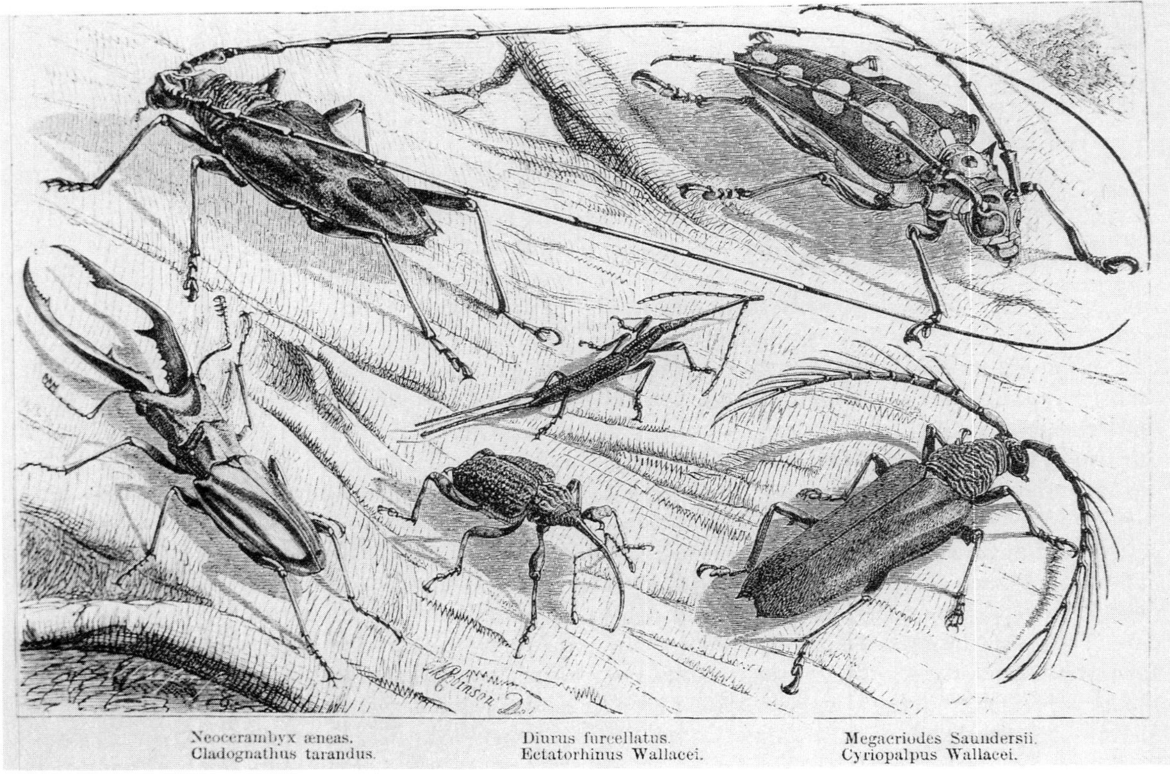

Fig. 10.1. 'Remarkable Bornean Beetles' collected by A. R. Wallace (*Malay Archipelago* 1869). Bottom left: a lamellicorn stag beetle, centre: two curculionid weevils.

terns. And, in addition, geometrical patterns of single taxa point up different similarities and dissimilarities between and within regions.

BIOGEOGRAPHY NOW

Pattern-making

The preceding chapters encompass both arithmetical and geometrical pattern-making. Sophisticated biological arithmetic has been used by van Balgooy and Holloway (Chapters 8 and 9) to plot the relationships of the islands of the Malay archipelago; geometric patterns of distribution of individual taxa are considered by Dransfield (Chapter 6).

Using Kroeber's (1916) measure of floristic affinity and cluster analysis, van Balgooy has measured the relatedness of the flora of Sulawesi to other islands (Table 8.1). He finds that, at the species level, Sulawesi is most closely related to the Moluccas, the Lesser Sunda Islands, and Java. The Sulawesi–Sunda block is then related to the Philippines. Least like Sulawesi is the Malaya–Sumatra–Borneo block. The two moth families analysed by Holloway are in striking agreement with the flowering plants. The close association of some moths and plants may exaggerate the similarities but, nevertheless, Holloway finds Sulawesi clustered with the Philippines and then with the Sunda islands. Both here and in an earlier study (Holloway and Jardine 1968) butterflies and birds also clustered Sulawesi with the Philippines (Fig. 9.1).

Table 10.1

Indices calculated according to Otsuka's formula to show family similarities between the Oriental and Australian regions in eight groups of animals. (Similarity = $\frac{C \times 100}{\sqrt{N_1 N_2}}$ where C is the number of shared taxa, N_1 the number in the larger fauna and N_2 the number in the smaller.)

	Similarity of Oriental to Australian	*Region of greatest similarity to Oriental*	*Region of greatest similarity to Australian*	*Category of organism*
Mammals (after Laurie and Hill 1954; Corbet and Hill 1980)	26	Ethiopian 71	Oriental 26	young immobile
Birds (after Gruson 1976, Howard and Moore 1984)	74	Palaearctic 86	Oriental 74	young mobile
Reptiles (after Darlington 1957)	64	Palaearctic Ethiopian 81	Palaearctic 69	young mobile
Amphibia (after Darlington 1957, Inger 1966)	35	Ethiopian 67	Neotropical 63	immobile
Land snails (after Solem 1979, 1981)	60	Australian 60	Oriental 60	old immobile
Dermaptera—earwigs (after Popham and Brindle 1966–69)	100	Australian Neotropical 100	Oriental Neotropical 100	old semi-mobile
Earthworms (after Sims 1978)	83	Australian 83	Oriental 83	old mobile
Freshwater flatworms (after Ball 1974)	67	Ethiopian 83	Neotropical 100	old immobile

The use of species as the unit of biotic diversity has disadvantages for biogeographical studies in Malesia. A fragmented island region provides ideal conditions for rapid and independent speciation. Island areas influence the number of species that can be accommodated (MacArthur and Wilson 1967). Altitude, climate, and even the length of the coastline influence speciation. Island species confuse the biologist by difficult-to-identify parallel evolution. It is, therefore, interesting to find that the genera of plants in van Balgooy's study (Table 8.2) show the same island relationships as the species: a Philippine–Sulawesi–Java–Lesser Sunda Islands block which relates closely to the Moluccas.

On the world scale, it has been the practice of zoologists to use the family as the unit of distribution. Familes may reveal more long-standing relationships than either species or genera. An analysis of the families of eight groups of animals, chosen for the availability of useful data, for their spread through the phylogenetic tree and for the mix of mobile and immobile taxa, can be used to compare the Oriental, Wallacean, and Australian regions. Several indices of similarity and dissimilarity exist and, although their calculation provides different sets of values, the hierarchical results are always the same. The important criterion of an index of similarity is that it should take into account the size of both faunas under

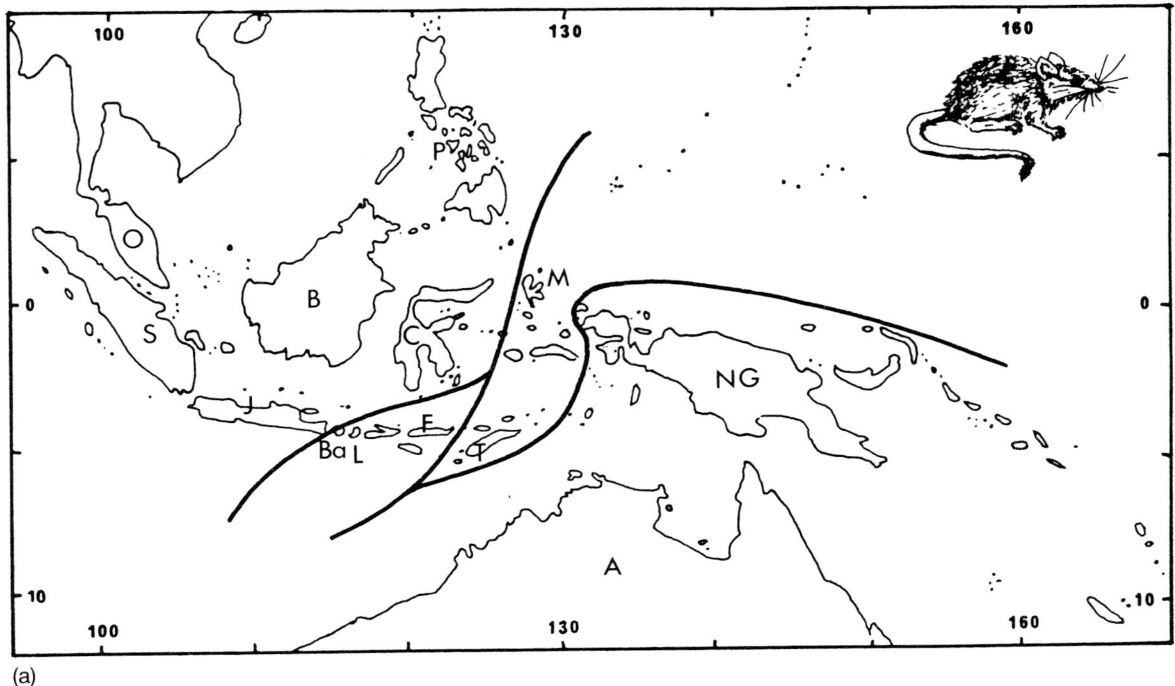

(a)

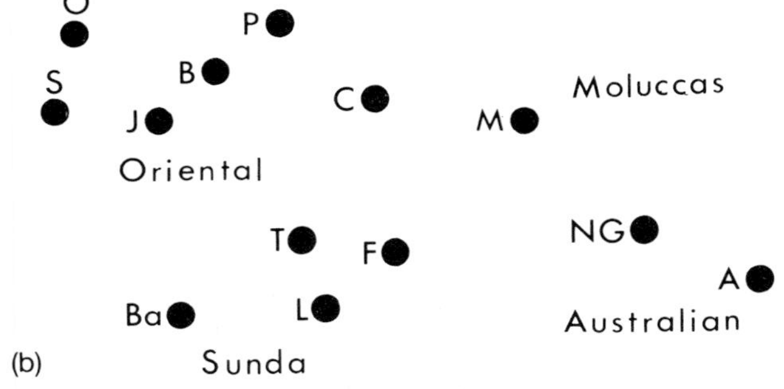

(b)

Fig. 10.2.
(a) map showing the main discontinuities of mammal families in the Malay archipelago; (b) relationships of the islands based on similarity indices of mammal families. The islands are heterogeneous, clustering into Oriental, Sunda, Molucca, and Australian groups. A: Australia, B: Borneo, Ba: Bali, C: Sulawesi, F: Flores, J: Java, L: Lombok, M: Moluccas, NG: New Guinea, P: Philippines, O: Oriental region, S: Sumatra.

consideration and also the shared taxa. In this way, both the total spread of fauna and the absence as well as presence of taxa is accounted for. For this reason Otsuka's easily calculated index (Cheetham and Hazel 1969) has been used to make Figs 10.2–10.5. (The highly sophisticated index of Preston (1962) used by Holloway and Jardine (1968) was rejected for the same reason that Simpson rejected it (1977): it requires complicated calculations and gives the same ordering of regions as the simple indices.) Table 10.1 shows the results for the Oriental and Australian regions. The main interest of the figures is the resemblance between the Oriental and Australian regions, whether the animals are phylogenetically young or old and whether they are comparatively mobile or immobile. The Oriental region also shows affinity with its

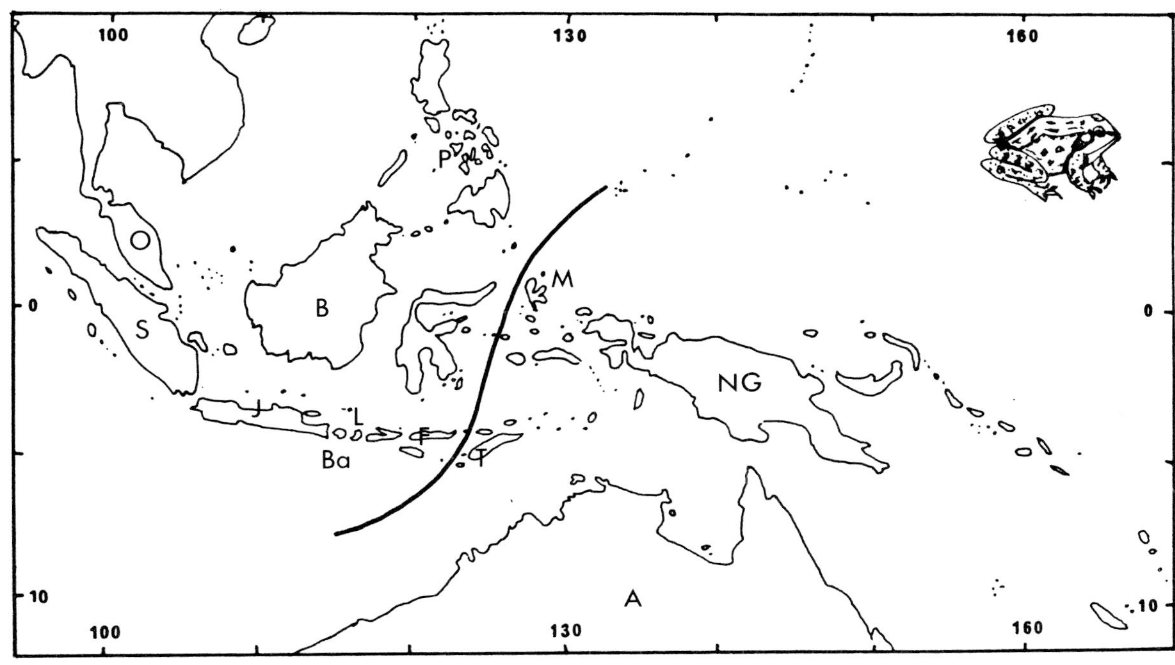

(a)

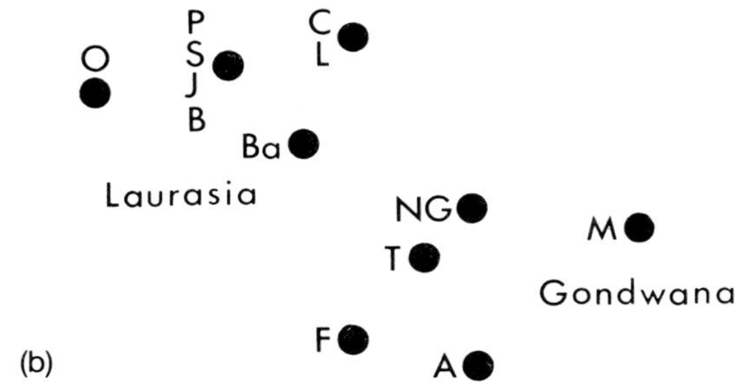

(b)

Fig. 10.3. (1) map showing the main discontinuities of amphibian families in the Malay archipelago; (b) relationships of the islands based on similarity indices of amphibian families. The islands cluster into a Laurasian and a Gondwana group. Abbreviations as in Fig. 10.2.

neighbours the Ethiopian and Palaearctic regions particularly in its 'young' vertebrate taxa, while the Australian region has close affinities with the Neotropical in some of the old immobile groups.

Applying the same calculations to the islands of Malesia, four taxa maps can be made (Figs 10.2–10.5) that can be equated with the bird map of Holloway and Jardine (1968), the butterflies

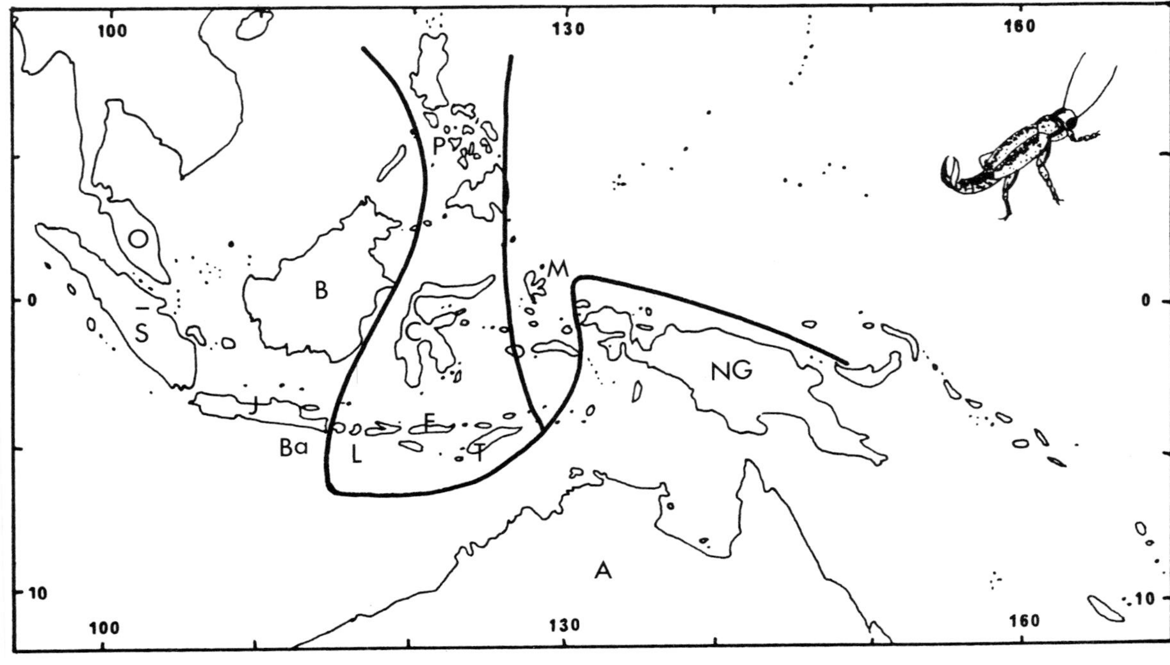

(a)

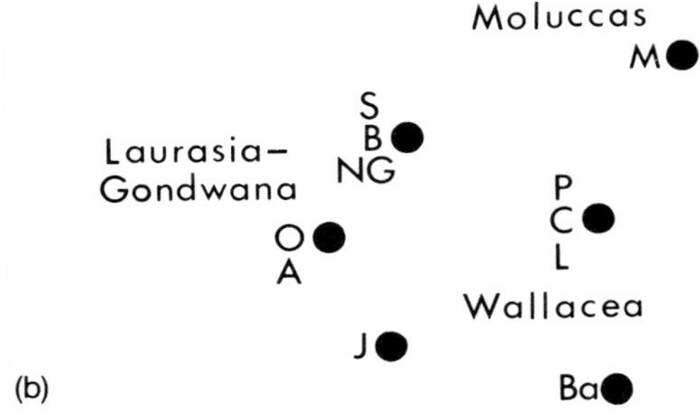

(b)

Fig. 10.4. (a) map showing the main discontinuities of earwig families in the Malay archipelago; (b) relationships of the islands based on similarity indices of earwig families. The islands cluster closely into Laurasian–Gondwana, Wallacean and Moluccan groups. (Timor and Flores are not included for lack of information.) Abbreviations as in Fig. 10.2.

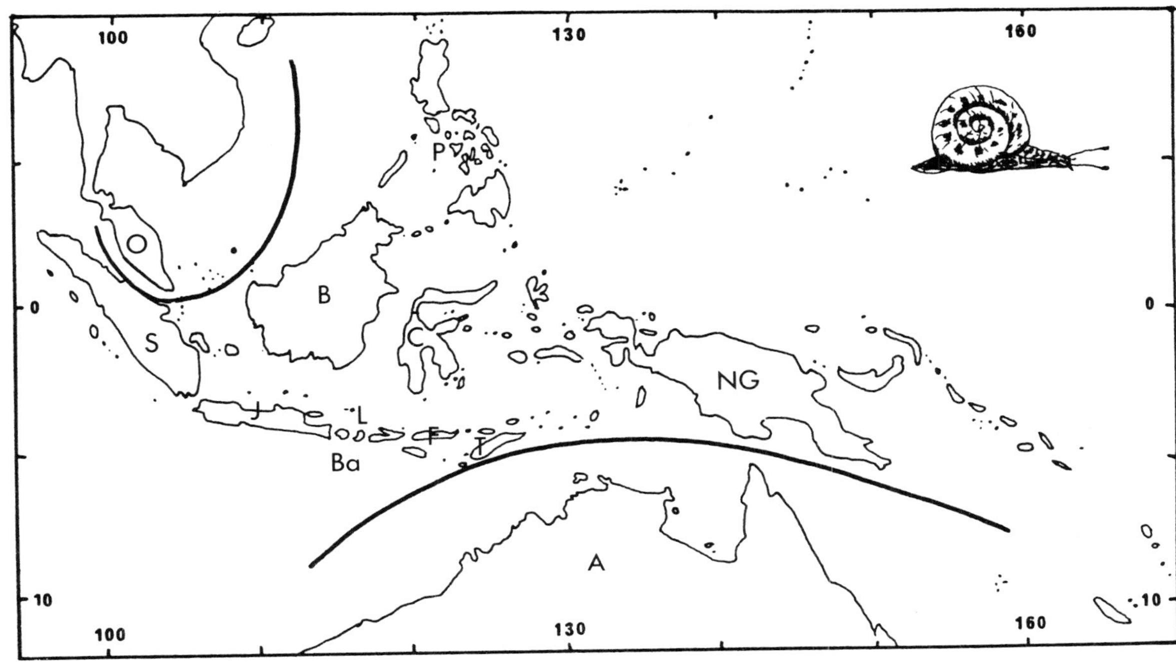

(a)

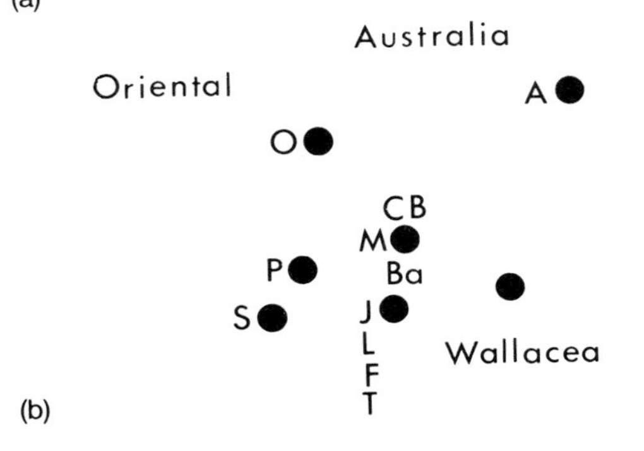

(b)

Fig. 10.5. (a) map showing the main discontinuities of land snail families in the Malay archipelago; (b) relationships of the islands based on similarity indices of land snail families. The islands cluster closely into Oriental, Wallacean and Australian groups. Abbreviations as in Fig. 10.2.

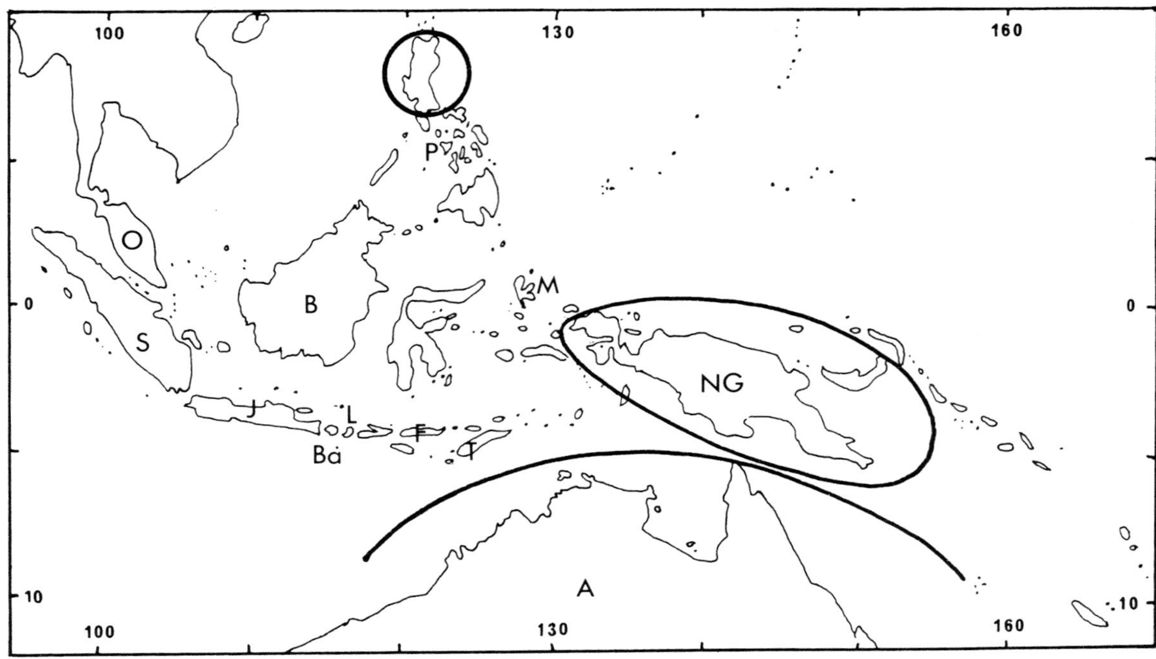

Fig. 10.6. The ranges of hydromyine 'water rat' genera in the Malay archipelago.

and moths of Holloway (Chapter 9) and the plants of van Balgooy (Chapter 8). In every case, except that of land snails and amphibia, Sulawesi most closely resembles the Philippines and the Lesser Sunda Islands; Sumatra and Borneo are close to one another and tie up with Java and the Oriental region. New Guinea is most like Australia with an occasional link with the Moluccas. Only among snails is there an anomalous tie-up between Sulawesi, Borneo, and the Moluccas and between New Guinea and the Lesser Sunda Islands, but the overall diversity among the snail faunas is small compared, for example, with the mammal faunas (Figs 10.2–10.5) where every island is unique. Across a wide spread of animals and plants, using species, genera, and families as 0the taxon unit, the results are broadly similar.

The geometrical pattern-makers in this volume have drawn attention to many enigmatic distributions in Malesia. Morley and Flenley (Chapter 5) attribute some patchy distributions to climatic conditions among the islands and Dransfield attributes the fragmented palm genus *Livistona* to the same cause. Dransfield finds anomalies in some of the bicentric patterns where there is no representative on at least some of the Malesian islands. Dransfield (1981) had already shown a major disjunction of the palms east of Borneo and west of New Guinea and in this volume he considers among others the peculiar distribution of the Calamoideae centred on Sundaland but with two genera, *Metroxylon* and *Pigafetta* in eastern Malesia. Some animals, too, exhibit bimodal distribution patterns in Malesia. Hydromyine 'water rats' (Fig. 10.6) have three genera in the Philippines, ten in New Guinea and two in Australia (Laurie and Hill 1954, Corbet and Hill 1980). Whitmore (1981) gave examples of a different pattern in which primarily southern and primarily northern pairs of families overlap in Malesia. The northern Ericaceae (heaths) and southern Epacridaceae (Australian heaths) do this. Among tree frogs (Fig. 10.7), the Old World

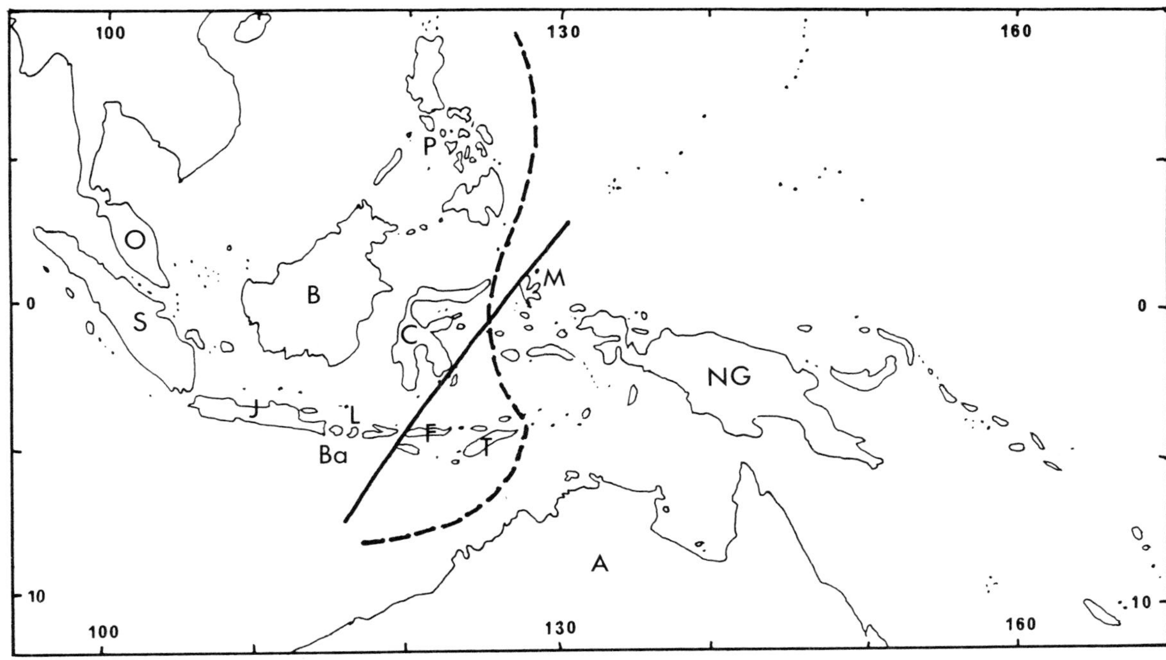

Fig. 10.7. The ranges of rhacophorid (broken line) and hylid (heavy line) tree frogs in the Malay archipelago.

tropic rhacophorids and the southern hylids meet in Malesia, only marginally crossing Wallace's line of 1910 (George 1981) and overlapping on Flores and Timor. Other patterns show mammal and snail (Solem 1981) families, for example, extending among the islands to different extents (Fig. 10.8).

Centres of origin theory

Both arithmetic and geometrical patterns indicate the abundance and diversity of angiosperm species and genera in Malesia and the strange mix of animals, poor in some species such as freshwater fish (Cranbrook 1981), rich in others such as parrots. Do centres of origin emerge from such patterns? Is it useful to talk about centres of origin? Is the 'birthplace' of a group the area where the greatest diversity is found? Takhtajan (1969 and Chapter 3), arguing for early angiosperm radiation somewhere between Assam and Fiji and, probably, in south-east Asia, emphasized the great variety of angiosperms in the area. The greatest diversity of murid rats and some birds is also found in south-east Asia, while the large number of *Rattus* species on Sulawesi led Groves (1976) to suggest that the genus originated there. Musser (Chapter 7) has dealt a severe blow to this thought with an improved classification that reduces the number of *Rattus* species from 24 to four. He gives a different interpretation of the murids of Sulawesi. The greatest diversity of kingfishers and bee-eaters is in Malesia (Fry 1969, 1980) and that of tupaiid tree-shrews is in Borneo (Cranbrook 1981). Is Malesia then the birthplace of kingfishers and bee-eaters and Borneo the birthplace of tree-shrews?

Or, is the centre of origin the place where the most plesiomorphic (primitive) taxon is found today (Hennig 1966), always assuming that there is agreement on which *is* the plesiomorph. The abundance of plesiomorphic families like the Magnoliaceae and Winteraceae in south-east Asia was a further reason for Takhtajan to assign

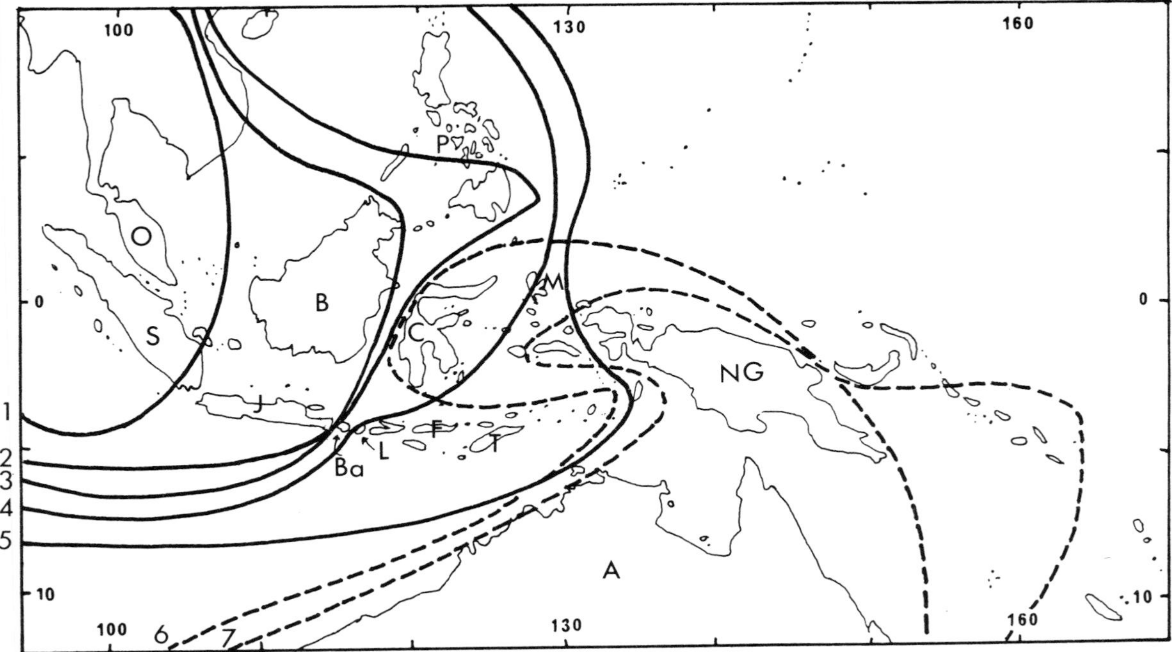

Fig. 10.8. Map showing differential extension of some Oriental and Australian mammal families through the Malay archipelago. 1: Tapiridae, 2: Rhinocerotidae, 3: Tupaiidae, 4: Suidae, 5: Soricidae, 6: Phalangeridae, 7: Peramelidae.

the birthplace of the angiosperms to the area. The most plesiomorphic genera of the murid rat and mouse family, thought to be *Apodemus* and *Micromys*, are, however, Palaearctic (Petter 1966, Chaline and Mein 1979).

Willis (1922) suggested the centre of origin was within the range of the most widespread species. Thus, the centre of origin of *Rhododendron* might be somewhere within the north Asian range of *R. anthopogon* or even within the circumpolar range of *R. lapponicum*. The most widespread species of tree-shrew *Tupaia glis* extends all over the Oriental region as far east as Java and Borneo.

The various hypotheses of centres of origin derived from modern biogeographical patterns do not provide concordant results. Did murids come from south-east Asia where they are most diverse or from the Palaearctic where the plesiomorphs live? Did *Rhododendron* originate in the Himalayas where the greatest number of species occurs or in a more northern area where the most widespread species exists today (Good 1947)? There are other explanations of diversity, like fragmentation of habitat and heterogeneity of climate. There is, for example, only one species of *Tupaia* on the Asian mainland but ten among the heterogeneous western islands of the Malay archipelago. There are, too, other explanations of plesiomorph location like relict populations. Plesiomorphic achatinellid tree snails are represented by relict populations on Pacific islands (Solem 1981) and *Ficus* species on the Solomon Islands are thought to be plesiomorphic relicts of a former widespread distribution (Corner 1967).

To add to the difficulties, not everyone agrees with current classifications and not all are based on acceptable cladistic identification of plesiomorphs.

Generalized tracks theory

The discrepancies between theories of centres of origin led to the emergence of the new theory of generalized tracks based on analysis of patterns (Croizat 1958). If the distributional tracks of diverse taxa coincide, then the pattern is said to have significance. The coincident tracks are considered to be indications of physical, geomorphological changes: vicariance. Animals and plants can only be moved about and isolated by only physical happenings. Generalized tracks often seem to imply a cosmopolitan origin for a taxon, a theory which, while getting away from obsessions with centres of origin, replaces them with an uninformative alternative. However, without the implications, generalized tracks are worth considering. In Malesia a very definite track for a diverse collection of taxa has been suggested in several chapters: a Philippine–Sulawesi–Sunda Island track. Already in 1939, Stresemann had identified such a track for the birds of the Malay archipelago; amphibians, butterflies, moths, earwigs, plants and perhaps even mammals can now be added. What does it mean? What do any of the patterns mean? Do they mean the same thing for all taxa? Do they provide evidence of the history of a taxon or the history of an area? Humboldt thought not. 'The distribution of organic beings over the surface of the earth does not depend solely on the great complications of thermic and climatic relations, but also on geological causes which continue almost wholly unknown to us' (Humboldt 1849).

Palaeontological evidence

By the late nineteenth century, geology and palaeontology were becoming less unknown subjects. The idea of a changing earth with evolving plants and animals, continuous over long periods of time, was the basis for a new biogeography. In *The geographical distribution of animals* (1876) Wallace defined two categories of study: zoological geography or pattern-making and geographical zoology which was 'the outcome of all preceding changes of the Earth'. Wallace believed that because both organisms and the land had changed over the millennia, fossils should play an important part in interpreting distribution patterns. Many of the pattern-makers disagree. They point out that fossils are worse than useless because a particular fossil is not necessarily the first of its kind nor at the centre of origin of its species and the point is made by Takhtajan (Chapter 3) in the case of the earliest angiosperm pollen. But, in spite of these limitations, the fossil record does establish that a particular organism was in a particular place at a particular time. To ignore this evidence in biogeography removes a dimension from the subject. Distribution patterns, reliable classification, palaeontology and geology all contribute to biogeography.

Recently, botanists have been finding important deposits of fossilized pollen. Truswell, Kershaw, and Sluiter (Chapter 5), reviewing the palynological finds in Australia, have argued that monosulcate angiosperms were comparatively late arrivals in Australia, appearing 10 Ma later than in Laurasia and Africa. Australia, they suggest, received late Cretaceous plants direct from the Gondwana coastal margin of Tethys. They do not agree with Takhtajan that angiosperms originated in 'a cradle somewhere between Assam and Fiji' and migrated round the world from there. Furthermore, they argue, fossil pollen finds do not agree with Dransfield's suggestion, from pattern analysis, that the rattan palm *Calamus* was an immigrant to New Guinea and Australia from south-east Asia in the later stages of the Tertiary because Truswell *et al.* believe they have pollen of *Calamus* from late Eocene deposits in south-east Australia.

Other angiosperm pollen suggests that Australia might have received very early Tertiary immigrants from the north like Myrtaceae (myrtles) as well as, perhaps, sending north some of its own products like Casuarinaceae (she oaks). This, and later interchanges, might be one of the causes of the present Malesian diversity.

More recent geomorphological changes have been outlined by Morley and Flenley (Chapter 5), studying like Humboldt, the distribution of mountain floras. But unlike Humboldt, they concentrate on change through time. From a distribution of pollen at different altitudes from early Pleistocene days it seems likely that land connections in the Malay archipelago have been greater than they are today and the climate cooler and drier. Thus, plants which today are isolated on mountain tops had an opportunity to migrate through lowlands when the climate was cooler. Plants of seasonal climates, at present disjunct in north and south Malesia, migrated, they suggest, through a possible savanna corridor from Thailand through Malaya and Java to Sulawesi.

Geological evidence

A complex picture of the origins of the flora and fauna of Malesia emerges and it may be the combined pressure of phytogeographers, zoogeographers and palynologists that stimulated Audley-Charles to further geological work in the area of south-east Asia. The reconstruction of the south-east Asian and Australian plates colliding round about Sulawesi in the mid-Miocene 15 Ma ago (which he reviewed in 1981) has been widely accepted.

Today, the story has become even more complex. Pieces of the northern margin of east Gondwana seem to have broken free in Jurassic days and rafted north as a Noah's ark (Pigram and Panggabean 1984; Audley-Charles, Chapter 2). Eventually, Audley-Charles thinks they collided with Laurasia to the west of Indo-China making the westerly part of south-east Asia a Gondwana–Laurasia mosaic with embedded terranes like those described for western North America. Because of the directions of movement, the northward drift, and the rotation of both Australia and the fragments, there was probably a land or island route from Laurasia to Australia in the late Cretaceous and through the Tertiary (Figs 2.6–2.9).

THE STORY SO FAR

With the combined evidence of the pattern-makers, the palaeontologists, and the geologists a new story of the Malay archipelago is being written. The story begins in the Jurassic with the northern edge of east Gondwana (which formed the southern shores of Tethys) breaking away as a number of blocks which included present-day South Tibet, Burma, Malaya, and Sumatra. They sailed north to Laurasia. But what organisms did they carry? Truswell *et al.* are convinced from fossil evidence that late Jurassic is too early for angiosperms in east Gondwana, but Takhtajan envisages such isolated fragments as a possible angiosperm birthplace, and Dransfield entertains the possibility of various palms being transported on the fragments, or the seeds using them as stepping-stones. Ancient wingless earwigs of Gondwana, found today in Java, the Oriental and Australian regions, but not on any other Malesian islands (Popham and Brindle 1966), might have been rafted on such fragments. The modern distribution of archaic helicinid land snails in the Neotropical, Oriental, Australian, and Malesian regions might suggest that they, too, used these Noah's arks, but, in this case, fossil evidence provides a good reason for rejecting the hypothesis: helicinid snails are known from Palaeozoic North America (Solem 1979).

As the Noah's arks sailed north and rotated there was a chance for Australian Gondwana products to meet south-east Asia Gondwana products along a land or archipelago route. This agrees with the fossil pollen reported by Truswell *et al.* The general homogeneity of ancient land snail families on the islands of the Malay archipelago (Fig. 10.5) might also be indicative of such a land route. Plesiomorph flatworms from Gondwana (Ball 1974) might have crawled north from Australia on their way to widespread colonizations. There is little fossil evidence so far for a massive interchange and Truswell *et al.* regard the connection as a stepping-stone route rather than a freeway: some plants making the journey, others, like *Nothofagus*, not. There is no indication that either marsupials or hylid tree frogs made any excursions along the corridor although they were supposedly sailing north on the Australasian plate at the time.

By the Oligocene, volcanic activity must have inhibited even the toughest plants and animals from migrating although Audley-Charles surmised that there was a route free from volcanic ash (Fig. 2.8). Did the Oligocene *Babyrusa* of Sulawesi, reported by Musser, arrive from Asia by that open route?

In the Miocene, 15 Ma ago, the well-established collision between the western and eastern parts of Sulawesi took place. It is possible that it was by this means that phalangers came from their Australian–Gondwana home to meet in Sulawesi the babirusa from Asia. The two floras and faunas mixed on Sulawesi, losing the locational identity of their separate origins. Neither van Balgooy, Holloway, or Musser has been able to find significantly Australasian communities in south-east Sulawesi to contrast with Oriental communities in the north. The Miocene collision also brought close together Australasian and Oriental taxa that did not overlap, such as several of the mammal families (Fig. 10.8). Soon after, with less volcanic activity in the area, there was an intermittent flow of plants and animals among the new islands: northern and southern taxa met and overlapped like the Ericaceae and Epacridaceae, the Rhacophoridae and the Hylidae. The flow of plants, according to Truswell *et al.*, was predominantly from east to west: they find little change in the Australian flora since the Miocene collision but suggest that the Winteraceae might have migrated west from Australia at this time. According to Dransfield, however, the Asian *Korthalsia* palm went east and reached as far as New Guinea, and the main flow of animals was also in that direction. Few mammals made the whole journey to Australia but those phylogenetically young murid rats and mice travelled the route from mid-Miocene Asia and eventually did reach Australia. The shrews stopped short (Fig. 10.8).

Several million years later, in the Pleistocene, Morley and Flenley's evidence suggests a temporary lowering of sea level and a cooler, drier climate. Was there a savanna land route for the stegodons to lumber along from south-east Asia through Java to Sulawesi and the eastern Sunda islands? Perhaps during that period a Philippine–Sulawesi–Sunda island route was also passable. Birds, butterflies and plants may have scattered through Malesia: some, like the woodpeckers, as far as Sulawesi; others, like the pines, no further than the Philippines and Sumatra. The choice of route depends on how much weight is given to distribution patterns of modern biota and how much to palynological and palaeoclimatological findings.

The haphazard nature of the colonizations and the dearth of freshwater fish east of Wallace's line (Cranbrook 1981) makes land connections dubious. Although there are so many possibilities for explaining the Malesian distributions by vicariant events there still remains a need to accept dispersal across physical barriers to explain, for example, the heterogeneity of the island mammal faunas (Fig. 10.2). Musser is strongly in favour of several dispersal events to account for the Sulawesi rats and mice. Taking advantage of temporarily congenial sea currents and winds, they spread in many directions.

There has been no discussion of the role of extinctions in the island pattern-making, although islands are notorious for plant and animal extinctions. Stegodons, giant pigs, and tortoises are known to have disappeared, but what others have gone without a trace? Is extinction responsible for some of the bicentric patterns?

The story is unfolding but there are still questions to be answered before it is complete.

BIBLIOGRAPHY

Achache, J. and Courtillot, V. (1984). Paleogeographic and tectonic evolution of southern Tibet since middle Cretaceous time: new paleomagnetic data and synthesis. *J. geophys. Res.* **89**, 10311–39.

Achache, J. and Courtillot, V. (1985). A preliminary Upper Triassic paleomagnetic pole for the Khorat plateau (Thailand): consequences for the accretion of Indochina against Eurasia. *Earth and Planetary Sci. Lett.* **73**, 147–57.

Albrecht, G. H. (1978). The craniofacial morphology of the Sulawesi macaques. *Contr. Primat.* 13, 1–151.

Allegre, C. J. and 34 others. (1984). Structure and evolution of the Himalaya-Tibet orogenic belt. *Nature, Lond.* **307**, 17–22.

Aleva, G. J. J. (1973). A contribution to the geology of part of the Indonesia tinbelt: the seas between Singkap and Banka islands and around the Karimata islands. (Proc. Reg. Conf. Geol. S.E. Asia). *Bull. Geol. Soc. Malaysia* **6**, 257–71.

Andel, T. H. van, Heath, G. R., Moore, T. C., and Mcgeary, D. F. R. (1967). Late Quaternary history, climate and oceanography of the Timor Sea, northwestern Australia. *Am. J. Sci.* **265**, 737–58.

Apandi, T. and Sudana, D. (1980). Geologic map of the Ternate Quadrangle, North Maluku 1:250,000. Geological Research and Development Centre of Indonesia.

Archbold, N. W., Pigram, C. J., Ratman, N., and Hakim, S. (1982). Permian brachiopod fauna from Irian Jaya, Indonesia: significance for Gondwana–south east Asia relationships. *Nature, Lond.* **296**, 556–8.

Ashton, P. S. (1972). The Quaternary geomorphological history of Western Malesia and lowland rain forest phytogeography. In *The Quaternary era in Malesia* (ed. P. and M. Ashton). Geography Department, University of Hull, Misc. Series 13.

Audley-Charles, M. G. (1974). Sulawesi. In *Mesozoic and Cenozoic organic belts. Data for orogenic studies* (ed. A. M. Spencer), pp. 365–78. Geological Society, London.

Audley-Charles, M. G. (1978). Indonesian and Philippine Archipelagoes. In *The Phanerozoic geology of the world II: The Mesozoic A* (ed. Moullade, M. and Nairn, A. E. M.), pp. 165–207. Elsevier, Amsterdam.

Audley-Charles, M. G. (1981). Geological history of the region of Wallace's line. In *Wallace's Line and plate tectonics* (ed. T. C. Whitmore). Clarendon Press, Oxford.

Audley-Charles, M. G. (1983). Reconstruction of eastern Gondwanaland. *Nature, Lond.* **306**, 48–50.

Audley-Charles, M. G. (1984). Cold Gondwana, warm Tethys and the Tibetan Lhasa block. *Nature, Lond.* **310**, 165–6.

Audley-Charles, M. G. (1985). The Sumba enigma: Is Sumba a diapiric nappe in process of formation? *Tectonophysics* **119**, 435–49.

Audley-Charles, M. G. and Hooijer, D. A. (1973). Relation of Pleistocene migrations of pygmy stegodonts to island arc tectonics in eastern Indonesia. *Nature, Lond.* **241**, 197–8.

Audley-Charles, M. G., Carter, D. J., Barber, A. J., Norvick, M. S., and Tjokrosapoetro, S. (1979). Reinterpretation of the geology of Seram: implications for the Banda Arcs and northern Australia. *J. geol. Soc. Lond.* **136**, 547–68.

Audley-Charles, M. G., Hurley, A. M., and Smith, A. C. (1981). Continental movements in the Mesozoic and Cenozoic. In *Wallace's Line and plate tectonics* (ed. T. C. Whitmore). Clarendon Press, Oxford.

Axelrod, D. I. (1959). Poleward migration of early angiosperm flora. *Science* **130**, 203–7.

Axelrod, D. I. (1961). How old are the angiosperms? *Amer. J. Sci.* **259**, 447–59.

Axelrod, D. I. (1970). Mesozoic paleogeography and early angiosperm history. *Bot. Rev.* **36**, 277–319.

Balgooy, M. M. J. van (1971). Plant geography of the Pacific, based on a census of Phanerogam genera. *Blumea, Suppl.* 6.

Ball, I. R. (1974). A continuation to the phylogeny and biogeography of the freshwater triclads (Platyhelminthes, Turbellaria). In *Biology of the Turbellaria* (ed. N. W. Riser and M. P. Morse), pp. 339–401. McGraw Hill, New York.

Bally, A. W. and 9 others (1980). Notes on the geology of Tibet and Adjacent areas—Report of the American plate tectonics delegation to the People's Republic of China. *US Geol. Survey Open File Report*, 80–501.

Barber, A. J., Audley-Charles, M. G., and Carter, D. J. (1977). Thrust tectonics in Timor. *J. geol. Soc. Austr.* **24**, 51–62.

Barlow, B. A. (1981). The Australian flora: its origin and evolution. In *Flora of Australia*, 1. Australian Government Printing Service, Canberra.

Barron, E. J., Harrison, C. G. A., Sloan, J. L., and Hay, W. W. (1981). Paleogeography, 180 million years ago to the present. *Eclogae geol. Helv.* **74**, 443–70.

Barstra, G. J. (1977). Walanae Formation and Walanae terraces in the Stratigraphy of South Sulawesi (Celebes, Indonesia). *Quartar* **27**/**28**, 21–30.

Barstra, G. J. (1978). Note on new data concerning the fossil vertebrates and stone tools in the Walanae Valley in South Sulawesi (Celebes). *Mod. Quat. Res. in Southeast Asia* **4**, 71–2.

Batchelor, B. C. (1979). Discontinuously rising late Cenzoic eustatic sea levels with special reference to Sundaland, Southeast Asia. *Geol. en Mijnbouw* **58**, 1–20.

Batten, D. J. (1984). Palynology, climate and the development of Late Cretaceous floral provinces in the Northern Hemisphere: a review. In *Fossils and Climate* (ed. P. Brenchley). Wiley, Chichester.

Beek, C. G. G. van (1982). *A Geomorphological and Pedological study of the Gunong Leuser National Park, North Sumatra, Indonesia*. Wageningen Agricultural University, Wageningen.

Bemmelen, R. W. van (1949). *The Geology of Indonesia*. Government Printing Office, The Hague.

Bergmans, W. (1978). On *Dobsonia* Palmer 1898 from the Lesser Sunda Islands (Mammalia: Megachiroptera). *Senckenbergiana biol.* **59**, 1–18.

Bergmans, W. (1979). Taxonomy and zoogeography of *Dobsonia* Palmer, 1898, from the Louisiade Archipelago, the D'Entrecasteaux Group, Trobriand Island and Woodlark Island (Mammalia, Megachiroptera). *Beaufortia* **355**, 199–214.

Bergmans, W. and F. G. Rozendaal (1981). Notes on Rhinolophus Lacepede, 1799 from Sulawesi, Indonesia, with the descriptions of new species (Mammalia, Michroptera). *Bijdragen tot de Dierkunde* **52**, 169–74.

Berry, E. W. (1911). Systematic paleontology Pteridophyta—Dicotyledoneae. In *Maryland geological survey, Lower Cretaceous* (ed. W. B. Clark). Baltimore.

Berry, E. W. (1926) *Cocos* and *Phymatocarpon* in the Pliocene of New Zealand. *Amer. J. Sci.* **212**, 181–4.

Biswas, B. (1973). Quaternary changes in sea-level in the South China Sea. (Proc. Reg. Conf. Geol. S.E. Asia). *Bull. Geol. Soc. Malaysia* **6**, 229–55.

Bowin, C., Purdy, G. M., Shor, G., Lawver, L., Hartono, H. M. S., and Jezek, P. (1977). Arc-continental collision in the Banda Sea region. *Bull. Am. Assoc. Petrol.*

Brenner, G. J. (1963). The spores and pollen of the Potomac Group of Maryland. *Maryland Dep. Geol. Mines Water Res.* **27**, 1–215.

Brenner, G. J. (1976). Middle Cretaceous floral provinces and early migration of angiosperms. In *Origin and early evolution of angiosperms* (ed. C. B. Beck). Columbia University Press, New York.

Brenner, G. J. (1984). Late Hauterivian angiosperm pollen from the Helez Formulation, Israel. *Sixth International Palynological Conference, Calgary, Abstracts*, 15.

Brown, M. and Earle, M. M. (1983). Cordierite-bearing schists and gneisses from Timor, eastern Indonesia: P-T conditions of metamorphism and tectonic implications. *J. metamorphic Geol.* **1**, 183–203.

Brunig, J. F. (1974). *Ecological studies in the Kerangas forests of Sarawak and Brunei*. Borneo literature Bureau, Kuching.

Buffetaut, E. (1981). Eléments pour une histoire paléobiogéographique du Sud-Est asiatique: l'apport des vertèbres fossils continentaux. *Bull. Soc. géol. France* (7), **XXIII**, 587–93.

Buffetaut, E. and Ingavat, R. (1980). A new corodilian from the Jurassic of Thailand, *Sunosuchus thailandicus* n. sp. (Mesosuchia, Goniopholididae), and the palaeogeographical history of south-east Asia in the Mesozoic. *Geobios* **13**(6), 879–89.

Buffetaut, E. and Ingavat, R. (1982). Phytosaur remains (reptilia, *Thecodontia*) from the upper Jurassic of north-eastern Thailand. *Geobios* **15**(1), 7–17.

Burbridge, N. T. (1960). The phytogeography of the Australian region. *Australian Journal of Botany*, **8**, 75–212.

Burger, D. (1980). Albian angiosperm distribution and habitat in Queensland, Australia. *Fifth International*

Palynological Conference, Cambridge, England, 1980. Abstracts, 65.

Burger, D. (1981). Observations of the earliest angiosperm development with special reference to Australia. *Proceedings, IV International Palynological Conference*, Lucknow (1976–77), **3**, 418–28.

Cameron, N. R., Clarke, M. C. G., Aldiss, D. T., Aspden, J. A., and Djunuddin, A. (1980). The geological evolution of northern Sumatra. 9th Ann. Convent. Indonesian Petrol. Assoc. Proc., 149–87.

Cande, S. C. and Mutter, J. C. (1982). A revised identification of the oldest sea-floor spreading anomalies between Australia and Antartica. *Earth and Planetary Science Letters* **58**, 151–60.

Candolle, A. de (1820). *Géographie botanique*. Levrault, Paris.

Carlquist, S. (1965). *Island life*. Natural History Press, New York.

Carlquist, S. (1974). *Island biology*. Columbia University Press, New York.

Carter, D. J., Audley-Charles, M. G., and Barber, A. J. (1976). Stratigraphical analysis of island arc-continental margin collision in eastern Indonesia. *J. geol. Soc. Lond.* **132**, 179–98.

Chaline, J. and Mein, P. (1979). *Les rongeurs et l'evolution*. Doin, Paris.

Cheetham, A. H. and Hazel, J. E. (1969). Binary (presence-absence) similarity coefficient. *Journal of Paleontology* **43**, 1130–6.

Chmura, C. A. (1973). Upper Cretaceous (Campanian–Maastrichtian) angiosperm pollen from the western San Joaquin Valley, California, U.S.A. *Palaeontographica* Abt B, **141**, 89–171.

Christophel, D. C. and Basinger, J. F. (1982). Earliest floral evidence for the Ebanaceae in Australia. *Nature, Lond.* **296**, 439–41.

Clason, A. T. (1976). A preliminary note about the animal remains from Ulu Liang I cave, South Sulawesi, Indonesia. *Mod. Quat. Res. in Southeast Asia* **2**, 53–67.

Climap Project Members (1976). The surface of the Ice-Age Earth. *Science, N.Y.* **191**, 1131–7.

Climap Project Members (1981). Seasonal reconstruction of the Earth's surface at the last glacial maximum. *Geol. Soc. America Map and Chart Series* MC-36.

Coetzee, J. A. (1981). A palynological record of very primitive angiosperms in Tertiary deposits of the south-western Cape Province, South Africa. *South Africa Journal of Science* **77**, 341–3.

Coetzee, J. A. and Muller, J. (1985). The phytogeographic significance of some extinct Gondwana pollen types from the Tertiary of the southwestern cape (South Africa). *Annals of the Missouri Botanical Garden* **71**, 1088–9.

Corbet, G. B. and Hill, J. E. (1980). *A world list of mammalian species*. British Museum (Natural History), London.

Corner, E. J. H. (1967). *Ficus* in the Solomon Islands and its bearing on the post-jurassic history of Melanesia. *Philosophical Transaction of the Royal Society* B **253**, 23–159.

Corner, E. J. H. (1976). *The seeds of dicotyledons*. I–II. Cambridge University Press.

Cornet, B. (1980). Tropical late Triassic monosulcate and polyaperture angiospermid pollen and their morphological relationship with associated auriculate polyplicate pollen. *Abstract. 5th Intern. Palynol. Conf. Cambridge*.

Cornet, B. (1981). Recognition of pre-Cretaceous angiosperm pollen and its relationship to fossil polyplicate pollen. *Abstract Proced. 12th Annual Meeting A.A.S.P. Palynol.* **5**, 212–13.

Couper, R. A. (1958). British Mesozoic microspores and pollen grains. *Palaeontographica*, Abt. B. **103**, 75–179.

Cracraft, J. (1980). Biogeography patterns of terrestrial vertebrates in the Southwest Pacific. *Palaeogeogr., Palaeoclimat., Palaeocol.* **31**, 353–69.

Cramwell, L. M., Carrington, H. J., and Speden, I. G. (1960). Lower Tertiary microfossils from McMurdo Sound, Antartica. *Nature, Lond.* **186**, 700–2.

Cranbrook, E. O. (1981). The vertebrate faunas. In *Wallace's Line and plate tectonics* (ed. T. C. Whitmore), Clarendon Press, Oxford.

Craw, R. C. and Weston, P. (1984). Panbiogeography: a progressive research program? *Syst. Zool.* **33**, 1–13.

Croizat, L. (1958). *Panbiogeography*. Vols 1, 2a and 2b. Caracas: the author.

Crostella, A. (1977). Geosynclines and plate tectonics in eastern Indonesia. *Am. Assoc. Petrol. Geol. Bull.* **61**, 2063–81.

Daghlian, C. (1981). A review of the fossil record of monocotyledons. *Bot. Rev.* **47**, 517–55.

Darlington, P. J. (1965) *Biogeography of the southern end of the world.* Harvard University Press.

Darlington, P. J. (1957). *Zoogeography: the geographical distribution of animals*. Wiley, New York.

Delson, E. (1980). Fossil macaques, phyletic relationships and a scenario of deployment. In *The macaques: studies in ecology, behaviour, and evolu-*

tion (ed. D. G. Lindburg), pp. 10–30. Van Nostrand Reinhold, New York.

Dermitzakis, M. D. and P. Y. Sondaar (1979). The importance of fossil mammals in reconstructing paleogeography with special reference to the Pleistocene Aegean Archipelago. *Ann. Geol. Helleniques* **29**, 808–40.

Dettmann, M. C. (1973). Angiospermous pollen from Albian to Turonian sediments of eastern Australia. *Geological Society of Australia*, Special Publication 4, 3–34.

Dettmann, M. E. (1981). The Cretaceous flora. In *Ecological biogeography in Australia* (ed. A. Keast). W. Junk, The Hague.

Dettmann, M. E. and Playford, G. (1969). Palynology of the Australian Cretaceous: a review. In *Stratigraphy and palaeontology: Essays in honour of Dorothy Hill* (ed. K. S. W. Campbell). ANU Press, Canberra.

Doutch, H. F.(1972). The palaeogeography of northern Australia and New Guinea and its relevance to the Torres Strait Area. In *Bridge and barrier: the natural and cultural history of Torres Strait* (ed. D. Walker). Australian National University, Canberra, Publication BG/3.

Doyle, J. A. (1969). Cretaceous angiosperm pollen of the Atlantic Coastal Plain and its evolutionary significance. *J. Arnold Arbor.* **50**, 1–35.

Doyle, J. A. (1973). Fossil evidence on early evolution of the monocotyledons. *Quart. Rev. Biol.* **48**, 399–413.

Doyle, J. A. (1977). Patterns of evolution in early angiosperms. In *Patterns of evolution as illustrated by the fossil record* (ed. A. Hallam). Developments in Palaeontology and Stratigraphy, 5. Elsevier, Amsterdam.

Doyle, J. A. (1978). Origin of angiosperms. *Ann. Rev. Ecol. Syst.* **9**, 365–92.

Doyle, J. A. (1984). Evolutionary, geographic, and ecological aspects of the rise of angiosperms. *Proceedings of the 27th International Geological Congress*, Vol. 2, pp. 23–33. VNU Science Press, Moscow.

Doyle, J. A. and Hickey, L. J. (1976). Pollen and leaves from the mid-Cretaceous Potomac Group and their bearing on early angio-sperm evolution. In *Origin and early evolution of angiosperms* (ed. C. B. Beck). Columbia University Press, New York.

Doyle, J. A. and Robbins, E. I. (1977). Angiosperm pollen zonation of the continental Cretaceous of the Atlantic Coastal Plain and its application to deep wells in the Salisbury Embayment. *Palynology* **1**, 43–78.

Doyle, J. A., Van Campo, M., and Lugardon, B. (1975). Observations on exine structure of *Eucommodites* and Lower Cretaceous angiosperm pollen. *Pollen et Spores* **17**, 429–86.

Doyle, J. A., Biens, P., Doerenkamp, R. and Jardiné, S. (1977). Angiosperm pollen from pre-Albian Lower Cretaceous of Equatorial Africa. *Bull. Cent. Rech. Explor. Prod. Elf-Aquitaine* **1**, 451–73.

Dransfield, J. (1972). The genus *Borassodenron* (Palmae) in Malesia. *Reinwardtia* **8**, 351–63.

Dransfield, J. (1981). Palms and Wallace's Line. In *Wallace's Line and plate tectonics* (ed. T. C. Whitmore). Clarendon Press, Oxford.

Dransfield, J. and Uhl, N. W. (1983). *Wissmannia* (Palmae) reduced to *Livistona*. *Kew Bull.* **38**, 199–200.

Dransfield, J. and Uhl, N. W. (1984). *Halmoorea*, a new genus from Madagascar with notes on *Sindroa* and *Orania*. *Principes* **28**, 163–7.

Dransfield, J., Comber, J. B., and Smith, G. (1986). A synopsis of *Corybas* Salisb. (Orchidaceae) in West Malesia and Asia. *Kew Bull.* **41**, 575–613.

Dransfield, J., Flenley, J. R., King, S. M., Harkness, D. D., and Rapu, S. (1984). A recently extinct palm from Easter Island. *Nature, Lond.* **312**, 750–2.

Dransfield, J., Irvine, A. K., and Uhl, N.W. (1985). *Oraniopsis appendiculata*, a previously misunderstood Queensland palm. *Principes* **29**, 56–63.

Duffels, J. P. (1983). Distribution patterns of Oriental Cicadoidea (Homoptera) east of Wallace's Line and plate tectonics. *Geo Journal* **7**, 491–8.

Earle, M. M. (1983). Continental margin origin for Cretaceous radiolarian cherts in western Timor. *Nature, Lond.* **305**, 129–30.

Essig, F. B. (1980). The genus *Orania* Zipp. (Arecaceae) in New Guinea. *Lyonia* **1**, 211–33.

Feiler, A. (1977). Uber die utrasoezufusche Variation des Phalangers ursinus (Mammalia, Marsupialia). *Zool. Abh. Mus. Tierk. Dresden* **34**, 187–97.

Feiler, A. (1978*a*). Bemerkungen uber Phalanger de 'orientalis-Gruppe' nach Tate (1945) (Mammalia, Marsupialia). *Zool. Abh. Mus. Tierk. Dresden* **34**, 385–95.

Feiler, A. (1978*b*). Uber artliche Abgrenzung und innerartliche Ausformung bei Phalanger maculatus (Mammalia, Marsupialia, Phalangeridae). *Zool. Abh. Mus. Tierk. Dresden* **35**, 1–30.

Feiler, A. (1978*c*). Zur morphologischem Charakteristik des Phalanger celbensis (Mammalia, Marsupialia, Phalangeridae). *Zool. Abh. Mus. Tierk. Dresden* **35**, 161.

Ferguson, I. K. (in press). Observations on the variation in pollen morphology of Palmae and its significance. *Canad. J. Bot.*

Flannery, T., George, G., Archer, M., and Maynes G. (in press). The Interrelationships of the living phalangerids (Phalangeroidea: Marsupialia) with a suggested new taxonomy.

Flenley, J. R. (1984). Later Quaternary changes of vegetation and climate in the Malesian Mountains. *Erdwissenschaftliche Forschung* **18**, 261–7.

Flenley, J. R. (1985). Quaternary vegetational and climatic history of Island South-East Asia. *Mod. Quat. Res. in Southeast Asia* **9**, 55–63.

Flenley, J. R. and Morley, R. J. (1978). A minimum age for the deglaciation of Mt. Kinabalu, East Malaysia. *Mod. Quat. Res. in Southeast Asia* **4**, 57–61.

Fontaine, W. M. (1889). The Potomac or Younger Mesozoic flora. *US Geol. Surv. Monogr.* **15**, 1–375.

Fooden, J. (1969). Taxonomy and evolution of the monkeys of the Celebes (Primates: Cercopithecidae). *Bibliotheca primatol.* **10**, 1–148.

Fooden, J. (1980). Classification and distribution of living macaques (Macaca Lacepede, 1799). In *The macaques: studies in ecology, behaviour, and evolution* (ed. D. G. Lindburg), pp. 1–9. Van Nostrand Reinhold, New York.

Fry, H. C. (1969). The evolution and systematics of bee-eaters (*Meropidae*). *Ibis* **111**, 557–92.

Fry, H. C. (1980).The evolutionary biology of the kingfishers (*Alcedinidae*). *Living Bird* **18**, 113–60.

George, W. (1981). Wallace and his line. In *Wallace's Line and plate tectonics* (ed. T. C. Whitmore), Clarendon Press, Oxford.

Girardeau, J. and 7 others (1984). Tectonic environment and geodynamic significance of the Neo-Cimmerian Donqiao ophiolite, Bangong-Nujiang suture zone, Tibet. *Nature, Lond.* **307**, 27–31.

Glover, I. C. (1981). Leang Burung 2: an Upper Palaeolithic rock shelter in South Sulawesi, Indonesia. *Mod. Quat. Res. in Southeast Asia* **6**, 1–38.

Gobbett, D. J. and Hutchison, C. S. (1973). *Geology of the Malay Peninsula*. Wiley-Interscience, New York.

Good, R. (1974). *The Geography of the flowering plants*. Longman, London.

Gould, R. E. (1976). The succession of Australian pre-Tertiary megafossil floras. *Botanical Review* **41**, 453–83.

Gould, S. J. (1977). *Ontogeny and phylogeny*. Harvard Univ. Press, Cambridge, Mass.

Groves, C. P. (1969). Systematics of the Anoa (Mammalia, Bovidae). *Beaufortia* **17**, 1–12.

Groves, C. P. (1976). The origin of the mammalian fauna of Sulawesi (Celebes). *Z. Saugetierk.* **41**, 201–16.

Groves, C. P. (1980*a*). Notes on the systematics of Babyrousa (*Artiodactyla, Suidae*). *Zool. Med., Leiden* **55**, 29–46.

Groves, C. P. (1980*b*). Speciation on Macaca; the view from Sulawesi. In *The macaques: studies in ecology, behaviour, and evolution* (ed. D. G. Londburg), pp. 84–124. Van Nostrand Reinhold, New York.

Groves, C. P. (1981). Ancestors for the pigs: taxonomy and phylogeny of the genus *Sus. Tech. Bull 3, Dept. Prehist., Res. School Pacific. Stud., Aust. Nat. Univ.* **i–iii**, 1–96.

Gruezo, W. M. and Harries, H. C. (1984). Self-sown, wild-type coconuts in the Philippines. *Biotropica* **16**, 140–7.

Gruson, E. S. (1976). *A checklist of the birds of the world.* Collins, London.

Hahn, L. and Siebenhuner, M. (1982). Explanatory notes (Paleontology) on the geological maps of northern western Thailand, 1:250,000. *Bund. Geowiss. und Rohstoffe, Hanover.*

Haile, M. S. (1981). Paleomagnetism of southeast and east Asia. *Amer. Geophys. Union Geodynamics Series* **2**, 129–35.

Hallam, A. (1984). Pre-Quaternary sea-level changes. *Ann. Rev. Earth Planet. Sci.* **12**, 205–43.

Hamilton, W. (1974). Earthquake map of the Indonesian region. *US Geol. Survey Misc. Inv. Ser. Map*, 1–875–C.

Hamilton, W. (1977). Subduction in the Indonesian region. *Am. Geophys. Union Maurice Ewing Series* **1**, 15–31.

Hamilton, W. (1979). *Tectonics of the Indonesian region*. US Geol. Survey Prof. paper 1078.

Harland, W. B., Cox, A. V., Llewellyn, P. G., Picton, C. A. G., Smith, A. C., and Walters, R. (1982). *A geologic time scale*. Cambridge University Press.

Harries, H. C. (1979). The evolution, dissemination and classification of *Cocos nucifera* L. *Botanical Rev.* **44**, 265–320.

Harries, H. C. (1984). Self-sown, wild-type coconuts from Australia. *Biotropica* **16**, 148–51.

Heaney, L. R. Species richness, dispersal, and vicariance of mammals in the Philippine Archipelago: a review of the problems. (Manuscript.)

Hekel, H. (1972). Pollen and spore assemblages from Queensland Tertiary sediments. *Geological Survey of Queensland Publication* **355** (Palaeontological Paper 30), 1–33.

Helmcke, D. (1982). On the Variscan evolution of central mainland southeast Asia. *Earth Evolution Sciences*, **4**, 309–19.

Hennig, W. (1966). *Phylogenetic systematics*. University of Illinois Press, Urbana.

Herbert, D. A. (1932). The relationships of the Queensland flora. *Proceedings, Royal Society of Queensland*, **44**, 2–22.

Herbert, D. A. (1967). Ecological segregation and Australian phytogeographic elements. *Proceedings, Royal Society of Queensland*, **78**, 101–11.

Hickey, L. J. and Doyle, J. A. (1977). Early Cretaceous fossil evidence for angiosperm evolution. *Bot. Rev.* **43**, 2–104.

Hickey, L. J., West, R. M., Dawson, M. R., and Choi, D. K. (1983). Arctic terrestrial biota: Paleomagnetic evidence of age disparity with mid-northern latitudes during the late Cretaceous and early Tertiary. *Science* **221**, 1153–6.

Hill, J. E. (1983). Bats (Mammalia: Chiroptera) from Indo-Australia. *Bull. Brit. Mus. Nat. Hist. (Zool.)* **45**, 103–208.

Holloway, J. D. (1970). The biogeographical analysis of a transect sample of the moth fauna of Mt. Kinabalu, Sabah, using numerical methods. *Biol. J. Linn. Soc.* **2**, 259–86.

Holloway, J. D. (1973). The taxonomy of four groups of butterflies (Lepidoptera) in relation to general patterns of butterfly distribution in the Indo-Australian area. *Trans. R. ent. Soc. London* **125**, 125–76.

Holloway, J. D. (1977). The Lepidoptera of Norfolk Island, their biogeography and ecology. *Series Entomologica* 13. W. Funk, The Hague.

Holloway, J. D. (1979). A survey of the Lepidoptera, biogeography and ecology of New Caledonia. *Series Entomologica* 15. W. Funk, The Hague.

Holloway, J. D. (1982*a*). Taxonomic appendix. In *An introduction to the moths of South East Asia* (ed. H. S. Barlow), pp. 174–271. Kuala Lumpur: the author.

Holloway, J. D. (1982*b*). Mobile organisms in a geologically complex area: Lepidoptera in the Indo-Australian tropics. *Zool. J. Linn. Soc.* **76**, 353–73.

Holloway, J. D. (1983). The moths of Borneo: family Notodontidae. *Malayan Nature J.* **37**, 1–107.

Holloway, J. D. (1984). Lepidoptera and the Melanesian Arcs. In *Biogeography of the tropical Pacific* (ed. F. J. Radovsky, P. H. Raven and S. H. Sohmer). Bishop Museum Special Publication, 72.

Holloway, J. D. (1985). The moths of Borneo: Family Noctuidae: Subfamilies Euteliinae, Stictopterinae, Plusiinae, Panthoinae. *Malayan Nature J.* **38**, 157–317.

Holloway, J. D. (1986). The moths of Borneo: key to families; families Cossidae, Metarbelidae, Ratardidae, Dudgeoneidae, Epipyropidae and Limacodidae. *Malay. Nat. J.* **40**, 1–165.

Holloway, J. D. (in press). Lepidopteran faunas of high mountains in the Indo-Australian tropics. In *High altitude tropical biogeography* (ed. M. Monasterio and F. Vuilleumier). Oxford University Press, New York.

Holloway, J. D. and Bender, R. (1985). Further notes on the Notodontidae of Sumatra, with descriptions of seven new species. *Heterocera Sumatrana*, **5**, 102–11.

Holloway, J. D. and Jardine, N. (1968). Two approaches to zoogeography: a study based on the distributions of butterflies, birds and bats in the Indo-Australian area. *Proc. Linn. Soc. Lond.* **179**, 153–88.

Hooghiemstra, H. (1984). *Vegetational and climatic history of the high plain of Bogotà, Colombia: a continuous record of the last 3.5 million years*. Cramer, Liechtenstein.

Hooijer, D. A. (1948). Pleistocene vertebrate from Celebes. III. *Anoa depressicornis* (Smith) subsp., and *Babyrousa babyrussa beruensis* nov. subsp. *Proc. kon. Ned. Akad. van Wetensch., Amsterdam* **51**, 3–11.

Hooijer, D. A. (1950). Man and other mammals from Toalian sites in south-western Celebes. *Verh. kon. Nederl. Akad. van Wetensch.* **46**, 1–165.

Hooijer, D. A. (1954). Pleistocene vertebrates from Celebes. VIII. Dentition and skeleton of *Celebochoerus heekereni* Hooijer. *Zool. Verh. Leiden* **24**, 1–46.

Hooijer, D. A. (1958). The Pleistocene vertebrate fauna of Celebes. *Asian Persp.* **2**, 71–6.

Hooijer, D. A. (1969). Pleistocene vertebrates from

Celebes. XIII. *Sus celebensis* Muller and Schlegel, 1845. *Beaufortia* **222**, 215–18.

Hooijer, D. A. (1971). A giant land tortoise, *Geochelone atlas* (Falconer and Cautley), from the Pleistocene of Timor. I. *Koninkl. Nederl. Akad. van Wetensch., Amsterdam*, Series B **74**, 504–25.

Hooijer, D. A. (1972). Pleistocene vertebrates from Celebes. XIV. Additions to the *Archidiskodon-Celebocheorus* fauna. *Zool. Med., Leiden* **46**, 1–16.

Hooijer, D. A. (1975). Quaternary mammals west and east of Wallace's Line. *Neth. J. Zool.* **25**, 46–56.

Hooijer, D. A. (1981). What, if anything new, is *Stegodon sumbaensis* Sartono? *Mod. Quaternary Res. in Southeast Asia* **6**, 89–90.

Hooijer, D. A. (1982). The extinct giant land tortoise and the pygmy stegodont of Indonesia. *Mod. Quaternary Res. in Southeast Asia* **7**, 171–6.

Hooker, J. D. (1860). Introductory Essay. *Botany of the Antarctic voyage of H.M. discovery ships 'Erebus' and 'Terror' in the years 1839–1843*, III. *Flora Tasmaniae*. Reeve, London.

Hope, G. S. and Peterson, J. A. (1975). Glaciation and vegetation in the high New Guinea mountains. *Roy. Soc. New Zealand Bull.* **13**, 155–62.

Hope, G. S. and Peterson, J. A. (1976). Palaeoenvironments. In *The equatorial glaciers of New Guinea* (ed. G. S. Hope, J. A. Peterson and U. Radok). A. A. Balkema, Rotterdam.

Howard, R. and Moore, A. (1984). *A complete check-list of the birds of the world*. Macmillan, London.

Hughes, N. F. (1976). *Palaeobiology of angiosperm origins*. Cambridge University Press.

Hughes, N. (1977). Palaeo-succession of earliest angiosperm evolution. *Bot. Rev.* **43**, 105–27.

Hughes, N. F., Drewry, G. E., and Laing, J. F. (1979). Barremian earliest angiosperm pollen. *Palaeontology* **22**, 513–35.

Humboldt, A. von (1849). *Views of nature* 3rd edn. Bohn, London.

Humphries, C. J. (1981). Biogeographical methods and the southern beeches. In *The Evolving Biosphere* (ed. P. L. Forey), British Museum (Nat. Hist). Cambridge University Press.

Inger, R. F. (1966). Systematics and zoogeography of amphibia of Borneo. *Fieldiana: Zoology* **52**, 1–402.

Jacobson, R. S., Shor, G. G., Kieckheffer, R. M., and Purdy, G. M. (1978). Seismic refraction and reflection studies in the Timor–Aru trough system and Australian continental shelf. *Mem. Am. Ass. Petrol. Geol.* **29**, 209–22.

Jenkins, P. D. (1976). Variation in Eurasian shrews of the genus *Crocidura* (Insectivora: Soricidae). *Bull. Brit. Mus. Nat. Hist. (Zool.)*, **30** 271–309.

Jenkins, P. D. (1982). A discussion of Malayan and Indonesian shrews of the genus *Crocidura* (Insectivora: Soricidae). *Zool. Med., Leiden* **31**, 267–97.

Johnson, B. D. and Veevers, J. J. (1984). Oceanic palaeomagnetism. In *Phanerozoic earth history of Australia* (ed. J. J. Veevers), pp. 17–38. Clarendon Press, Oxford.

Johnson, L. A. S. and Briggs, B. G. (1975). On the Proteaceae, the evolution and classification of a southern family. *Bot. J. Linn. Soc.* **70**, 88–182.

Johnson, L. A. S. and Briggs, B. G. (1981). Three old southern families—Myrtaceae, Proteaceae and Restionaceae. In *Ecological biogeography of Australia* (ed. A. Keast), pp. 427–69. W. Junk, The Hague.

Jong, N. de and Bergmans, W. (1981). A revision of the fruit bats of the genus *Dobsonia* Palmer, 1898 from Sulawesi and some nearby islands (Mammalia, Megachiroptera, Pteropodinae). *Zool. Abh. Mus. Tierk. Dresden* **37**, 209–24.

Karig, D. E. (1971). Origin and development of marginal basins in the western Pacific. *J. geophys. Res.* **76**, 2542–61.

Katili, J. A. (1978). Past and present geotectonic position of Sulawesi, Indonesia. *Tectonophysics* **45**, 289–322.

Kemp, E. M. (1968). Probable angiosperm pollen from British Barremi to Albian strata. *Palaeontology* **11**, 421–34.

Kemp, E. M. (1978). Tertiary climatic evolution and vegetation history in the southeast Indian Ocean region. *Palaeogeography, Palaeoclimatology, Palaeoecology* **24**, 169–208.

Kemp, E. M. and Harris, W. K. (1977). The palynology of early Tertiary sediments, Ninetyeast Ridge, Indian Ocean. *Special Papers in Palaeontology* **19**, 1–70.

Kennett, J. P. and Watkins, N. D. (1974). Late Miocene–Early Pliocene palaeomagnetic stratigraphy, paleoclimatology and biostratigraphy in New Zealand. *Bull. Geol. Soc. Am.* **85**, 1380–98.

Kennett, J. P., Houtz, R. E., Andrews, P. B., Edwards, A. R., Gostin, V. A., Hajos, M., Hampton, M. A., Jenkins, D. G., Margolis, S. V., Ovenshire, A. T., and Perch-Nielsen, K. (1975). Cenozoic palaeoceanography in the southwest Pacific Ocean, Antarctic glaciation and the development of the circum Antarctic current. *Initial Reports Deep Sea Drilling Project* 29 (ed. J. P. Kennett, R. E. Houtz *et al.*). US Gov. Print. Office, Washington.

Kenyon, C. S. (1974). *Stratigraphy and sedimentology of the Late Miocene to Quaternary deposits of Timor* Ph.D. thesis, University of London.

Kershaw, A. P. (1970). A pollen diagram from Lake Euramoo, northeast Queensland, Australia. *New Phytologist* **69**, 785–805.

Kershaw, A. P. (1976). A Late Pleistocene and Holocene pollen diagram from Lynch's Crater, Northeastern Queensland, Australia. *New Phytol.* **77**, 469–98.

Kershaw, A. P. (1985). An extended late Quaternary vegetation record from north-eastern Queensland and its implications for the seasonal tropics of Australia. *Proceedings of the Ecological Society of Australia* 13.

Kershaw, A. P. and Sluiter, I. R. (1982). Late Cainozoic pollen spectra from the Atherton Tableland, northeastern Australia. *Australian Journal of Botany* **30**, 279–95.

Khan, A. M. (1974). Palynology of Neogene sediments from Papua (New Guinea) stratigraphic boundaries. *Pollen et Spores* **16**, 265–84.

Kiriakoff, S. G. (1967). New genera and species of Oriental Notodontidae (Lepidoptera). *Tijdschr. Ent.* **110**, 37–64.

Koenigswald, von G. H. R. (1967). An upper Eocene mammal of the family Anthracotheriidae from the island of Timor, Indonesia. *Koninkl. Nederl. Akad. Wetensch. Proc.* B-**70**, 528–33.

Komarov, V. L. (1908). Introduction to the floras of China and Mongolia. *Acta Horti Petropolitani* **29** (1), 1–176; **29** (2), 179–388. (In Russian).

Koning, R. de and Sosef, M. S. M. (in prep.). *The floristic position of Sulawesi*.

Koopman, K. F. (1979). Zoogeography of mammals from islands off the northeastern coast of New Guinea. *Amer. Mus. Novit.* **2690**, 1–17.

Koopman, K. F. (1982). Results of the Archbold Expeditions No. 109. Bats from eastern Papua and the East Papuan islands. *Amer. Mus. Novit.* **2747**, 1–34.

Koopman, K. F. (1984). Taxonomic and distributional notes on tropical Australian bats. *Amer. Mus. Novit.* **2778**, 1–48.

Koopmans, B. N. and Stauffer, P. H. (1968). Glacial phenomena on Mount Kinabalu, Sabah, Borneo region. *Malaysia Geological Survey Bull.* **8**, 25–35.

Kowal, N. E. (1966). Shifting cultivation, fire and pine forest in the Cordillera Central, Luzon, Philippines. *Ecol. Monographs* **36**, 389–419.

Krassilov, V. A. (1967). *The Early Cretaceous flora of Southern Primorie and their stratigraphic significance*. Nauka, Moscow. (In Russian.)

Kroeber, A. L. (1916). Floral relations among the Galapagos Islands, *Univ. Cal. Publ. Bot.* **6**, 199–220.

Kryshtofovich, A. N. (1929). Discovery of the oldest dicotyledons of Asia in the equivalents of the Potomac Group in Suchan Ussuriland, Siberia. *Bull. du Comité Géolog., Leningrad*, **48** (6–10), 1357–91. (In Russian with English summary.)

Lacordaire, J. T. (1839). On the geographical distribution of insects. *Edinburgh New Philosophical Journal* **27**, 170–83, 224–38.

Laing, J. F. (1976). The stratigraphic setting of early angiosperm pollen. In *The evolutionary significance of the exine* (ed. I. K. Ferguson and J. Muller), pp. 15–26. Academic Press, London.

Lam, H. J. (1945). Notes on the historical phytogeography of Celebes. *Blumea* **5**, 600–40.

Lapouille, A., Haryono, H., Larne, M., Pramumijoyo, S., and Lardy, M. (in press). Age and origin of the sea floor of the Banda Sea (eastern Indonesia). *Tectonophysics*.

Latreille, P. A. (1821). Introduction à la géographie générale des Aracnides et des Insectes. *Edinburgh New Philosophical Journal* **5**, 370–8.

Laurie, E. M. O. and Hill, J. E. (1954). *List of land-mammals of New Guinea, Celebes and adjacent islands*. British Museum (Natural History), London.

Löffler, E. (1984). Pleistocene and present day glaciations in the high mountains of New Guinea. *Erdwissenschaftliche Forschung* **18**, 249–59.

MacArthur, R. H. and Wilson, E.O. (1967). *The theory of island biogeography*. Princeton University Press.

MacKinnon, J. and MacKinnon, K. (1980). The behavior of wild spectral tarsiers. *Internat. Jour. Primatology* **1**, 361–79.

Maloney, B.K. (1980). Pollen analytical evidence for early forest clearance in North Sumatra. *Nature, Lond.* **287**, 324–6.

Maloney, B. K. (1981). A pollen diagram from Tao Sipinggan, a lake site in the Batak Highlands of North Sumatra, Indonesia. *Mod. Quat. Res. in Southeast Asia*, **6**, 57–66.

Martin, H. A. (1978). Evolution of the Australian flora and vegetation through the Tertiary: evidence from pollen. *Alcheringa* **2**, 181–202.

Martin, H. A. (1982). Changing Cenozoic barriers and

the Australian paleobotanical record. *Annals of the Missouri Botanical Garden* **69**, 625–67.

McCaffrey, R. (1983). Seismic wave propagation beneath the Molucca Sea arc–arc collision zone, Indonesia. *Tectonophysics* **96**, 45–57.

McKenna, M. C. (1973). Sweepstakes, filters, corridors, Noah's Arks, and beached viking funeral ships and palaeogeography. In *Implications of continental drift to the earth sciences* (ed. D. H. Tarling and S. K. Runcorn), pp. 293–308. Academic Press.

McTavish, R. A. (1975). Triassic conodonts and Gondwana stratigraphy. *Gondwana Geology* (ed. K. S. W. Campbell), pp. 481–90. A.N.U. Press, Canberra.

Medway, Lord (1972). The Quaternary mammals of Malesia. In *The Quaternary Era in Malesia* (ed. P. S. and M. Ashton). Geography Department, University of Hull, Misc. Series 13.

Meijden, R. van der (1982). Systematics and evolution of *Xanthophyllum*. *Leiden Bot. Ser.* 7.

Meijer, W. (1982). Plant refuges in the Indo-Malesian region. In *Biological diversification in the tropics* (ed. G. T. Prance), pp. 576–84. Columbia University Press, New York.

Meyer, A. B. (1896). Säugethiere vom Celebes- und Philippinen-Archipel. *Abhandlungen der zoologischen Anthropologisch-ethnologischen Museum, Dresden*, **6**, 1–36.

Meyer, A. B. (1898). Sängethiere vom Celebes und Philippinen Archipel. II. *Abhandlungen der zoologischen Anthropologisch-ethnologischen Museum, Dresden*, **7**, 1–55.

Mitchell, A. H. G. (1981). Phanerozoic plate boundaries in mainland S.E. Asia, the Himalayas and Tibet. *J. geol. Soc. Lond.* **138**, 109–22.

Mitchell, A. H. G. (in press). Mesozoic and Cenozoic regional tectonics and metallogenesis in mainland S.E. Asia. *Bull. geol. Soc. Malaya.*

Molnar, P. and Tapponnier, P. (1975). Cenozoic tectonics of Asia: effects of a continental collision. *Science* **189**, 419–26.

Moore, G. F. and Silver, E. A. (1982). Collision processes in the northern Molucca Sea. *Geophys. Monograph Series*, 27, 360–72.

Moore, G. F., Kadurisman, D. K., Evans, C. A., and Hawkins, J. W. (1981). Geology of the Talaud islands, Molucca Sea collision zone, northeast Indonesia. *J. Structural Geol.* **3**, 467–75.

Moore, H. E. Jr (1973*a*). The major groups of palms and their distribution. *Gentes Herb.* **11**, 27–141.

Moore, H. E. Jr (1973*b*). Palms in the tropical forest ecosystems of Africa and South America. In *Tropical forest ecosystems in Africa and South America: a comparative review* (ed. B. J. Meggers, E. S. Ayensu, and W. D. Duckworth). Smithsonian Institution Press, Washington.

Moore, H. E. Jr and Uhl, N. W. (1982). The major trends of evolution in palms. *Botanical Rev.* **48**, 1–69.

Moore, J. C. (1958). A new species and a redefinition of the squirrel genus *Prosciurillus* of Celebes. *Amer. Mus. Novit.* **1890**, 1–5.

Morley, R. J. (1978). Palynology of Tertiary and Quaternary sediments in Southeast Asia. *Proc. Indonesian Petroleum Ass. 6th Ann. Conv.* 1977.

Morley, R. J. (1981). Development and vegetation dynamics of a lowland ombogenous peat swamp in Kalimantan Tengah, Indonesia. *J. Biogeogr.* **8**, 383–404.

Morley, R. J. (1982). A palaeoecological interpretation of a 10,000 year pollen record from Danau Padang, central Sumatra, Indonesia. *J. Biogeogr.* **9**, 151–90.

Muller, J. (1966). Montane pollen from the Tertiary of N.W. Borneo. *Blumea* **14**, 231–5.

Muller, J. (1968). Palynology of the Pedawan and Plateau Sandstone Formations (Cretaceous-Eocene) in Sarawak, Malaysia. *Micropalaeontology* **14**, 1–37.

Muller, J. (1972). Palynological evidence for change in geomorphology, climate and vegetation in the Mio-Pliocene of Malesia. In *The Quaternary Era in Malesia* (eds. P. S. and M. Ashton). Geography Department, University of Hull, Misc. Series 13.

Muller, J. (1979). Reflections on fossil palm pollen. Proc. IV. *Int. Palyn. Conf. Lucknow* **1**, 568–79.

Muller, J. (1981). Fossil pollen records of extant angiosperms. *Botanical Review* **47**, 1–142.

Mulvaney, D. J., and Soejono, R. P. (1970). The Australian–Indonesian archeological expedition to Sulawesi. *Asian Perspect.* **13**, 163–77.

Musser, G. G. (1969). Results of the Archobold Expeditions. No. 91. A new genus and species of murid rodent from Celebes, with a discussion of its relationships. *Amer. Mus. Novit.* **2384**, 1–41.

Musser, G. G. (1981). The giant rat of Flores and its relatives east of Borneo and Bali. *Bull. Amer. Mus. Nat. Hist.* **169**, 67–176.

Musser, G. G. (1982). Results of the Archbold Expeditions. No. 110. *Crunomys* and the small-bodied shrew rats native to the Philippine Islands and Sulawesi (Celebes). *Bull. Amer. Mus. Nat. Hist.* **174**, 1–95).

Musser, G. G. (1984). Identities of subfossils rats from caves in southwestern Sulawesi. *Mod. Quaternary Res. in Southeast Asia* **8**, 61–94.

Musser, G. G. and Dagosto, M. *The Sulawesi tarsiers* (Manuscript.)

Musser, G. G. and Heaney, L. R. (1985). Philippine *Rattus*: A new species from the Sulu Archipelago. *Amer. Mus. Novit.* **2818**, 1–32.

Musser, G. G. and Newcomb, C. (1983). Malaysian murids and the giant rat of Sumatra. *Bull. Amer. Mus. Nat. Hist.* **174**, 327–598.

Musser, G. G., Heaney, L. R., and Rabor, D. S. (1985). Philippine rats: a new species of *Crateromys* from Dinagat Island. *Amer. Mus. Novit.* **2821**, 1–25.

Nelson, E. C. (1981). Phytogeography of southern Australia. In *Ecological biogeography of Australia* (ed. A. Keast), pp. 733–59. W. Junk, The Hague.

Niemitz, C. (1977). Zur Funktionsmophologie and Biometrie der Gattung *Tarsius* Storr, 1780. Herleitung von Evolutionsmechanismen bei einem Primaten. *Cour. Forsch.-Inst. Senckenberg* **25**, 1–161.

Niemitz, C. (1984*a*). An investigation and review of the territorial behaviour and social organization of the genus *Tarsius*. In *Biology of tarsiers* (ed. C. Niemitz), pp. 117–27. Gustav Fischer Verlag, Stuttgart and New York.

Niemitz, C. (1984*b*). Taxonomy and distribution of the genus Tarsius Storr, 1780. In *Biology of tarsiers* (ed. C. Niemitz), pp. 1–16. Gustav Fischer Verlag, Stuttgart and New York.

Niemitz, C. (1985). Der Koboldmaki. Evolutionsforschung an einem Primaten. *Naturwissenschaftliche Rundschau* **38**, 43–9.

Niklas, K. J., Tiffney, B. H., and Knoll, A. H. (1980). Apparent changes in the diversity of fossil plants. *Evolutionary Biology* **12**, 1–89.

Nishida, M. (1962). On some petrified plants from the Cretaceous of Chôshi, Chiba Prefecture. *Jap. J. Bot.* **18**, 87–104.

Norvick, M. S. (1979). The tectonic history of the Banda Arcs, eastern Indonesia, a review. *J. geol. Soc. Lond.* **136**, 519–27.

Norvick, M. S. and Burger, D. (1976). Palynology of the Cenomanian of Bathurst Island, Northern Territory, Australia. *Bulletin Bureau of Mineral Resources, Geology and Geophysics, Australia* **151**, 1–169.

Nur, A. and Ben-Avraham, Z. (1977). Lost Pacifica continent. *Nature, Lond.* **270**, 41–3.

Paltrinieri, F. and Saint-Marc, P. (1976). Stratigraphic and paleogeographic evolution during Cenozoic time in western Indonesia. *Seapex Program, Offshore South Asia Conference, February 1976*, Paper 10.

Parker, E. S. and Gealey, W. K. (1985). Plate tectonic evolution of the western Pacific–Indian Ocean region. *Energy* **10**, 249–61.

Parsons, B. and Sclater, J. G. (1977). An analysis of the variation of ocean floor bathymetry and heat flow with age. *J. Geophys. Res.* **82**, 803–827.

Petter, F. (1966). L'origine des muridés: plan cricétin et plans murins. *Mammalia* **30**, 204–25.

Phengklai, C. (1972). Pinaceae, Cephalotaxaceae, Cupressaceae. *Flora of Thailand* **2**, 193–6.

Pigram, C. J. and Panggabean, H. (1983). Age of the Banda Sea, eastern Indonesia. *Nature, Lond.* **301**, 231–4.

Pigram, C. J. and Panggabean, H. (1984). Rifting of the northern margin of the Australian continent and the origin of some micro-continents in eastern Indonesia. *Tectonophysics* **107**, 331–53.

Popham, E. J. and Brindle, A. (1966–69). Genera and species of the Dermaptera. *Entomologist* **99**, **100**, **101**, **102**.

Powell, C. M. and Johnson, B. D. (1980). Constraints on the Cenozoic position of Sundaland. *Tectonophysics* **63**, 91–109.

Powell, C. M., Johnson, B. D., and Veevers, J. J. (1981). The Early Cretaceous break-up of eastern Gondwanaland, the separation of Australia and India, and their interaction with southeast Asia. In *Ecological biogeography of Australia* (ed. A. Keast), pp. 15–29. W. Junk, The Hague.

Powell, D. E. (1976). The geological evolution and hydrocarbon potential of the continental margin off north-west Australia. *APEA J.* **16**(1), 12–23.

Preston, F. W. (1962). The canonical distribution of commonness and rarity. *Ecology* **43**, 185–215, 410–32.

Price, N. J. and Audley-Charles, M. G. (1983). Plate rupture by hydraulic fracture resulting in overthrusting. *Nature, Lond.* **306**, 572–5.

Quilty, P. G. (1984). Phanerozoic climates and environments of Australia. In *Phanerozoic earth history of Australia* (ed. J. J. Veevers), pp. 48–57. Clarendon Press, Oxford.

Raven, P. H. (1979). Plate tectonics and southern

hemisphere biogeography. In *Tropical botany* (ed. K. Larsen and L. B. Holm-Nielsen). Academic Press, London.

Raven, P. H. and Axelrod, D. I. (1972). Plate tectonics and Australasian palaeobiogeography. *Science* **176**, 1379–86.

Raven, P. H. and Axelrod, D. I. (1974). Angiosperm biogeography and past continental movements. *Ann. Miss. Bot. Gard.* **61**, 539–673.

Reiner, E. (1960). The glaciation of Mount Wilhelm, Australia New Guinea. *Geogrl. Rev.* **40**, 491–503.

Retallack, G. and Dilcher, D. L. (1981). A coastal hypothesis for the dispersal and rise to dominance of flowering plants. In *Paleobotany, paleoecology and evolution* (ed. K. J. Niklas), pp. 27–77. Praeger, New York.

Ridd, M. F., 1971. South-east Asia as a part of Gondwanaland. *Nature, Lond.* **234**, 531–3.

Rookmaaker, L. C. and Bergmans, W. (1981). Taxonomy and geography of *Rousettus amplexicaudatus* (Geoffrey, 1810) with comparative notes on sympatric congeners (Mammalia, Megachiroptera). *Beaufortia* **31**, 1–29.

Royen, P. van (1983). *The genus* Corybas *(Orchidaceae) in its eastern areas*. Phanerogamarum Monographiae Tomas XVI. J. Cramer, Vaduz.

Rozendaal, F. G. (1984). Notes on macroglossine bats from Sulawesi and the Moluccas, Indonesia, with the description of a new species of *Syconycteris* Matschie, 1899 from Halmahera (Mammalia: Megachiroptera). *Zool. Med., Leiden* **58**, 187–212.

Saporta, G. de (1894). *Flora fossile du Portugal*. Lisbon.

Sartono, S. (1973). On Pleistocene migration routes of vertebrate faunas in southeast Asia. *Bull. Geol. Soc. Malaysia* **6**, 273–86.

Sartono, S. (1979). The age of vertebrate fossils and artifacts from Cabenge in South Sulawesi, Indonesia. *Mod. Quat. Res. in Southeast Asia* **5**, 65–81.

Scholtz, A. (1985). The palynology of the upper lacustrine sediments of the Arnot Pipe, Banke, Namaqualand. *Annals of the South African Museum* **95**, 1–109.

Schuster, R. M. (1972). Continental movements, Wallace's Line and Indo–Malayan Australasian dispersal of land plants; some eclectic concepts. *Botanical Rev.* **38**, 3–86.

Schuster, R. M. (1976). Plate tectonics and its bearing on the geographical origin and dispersal of angiosperms. In *Origin and early evolution of angiosperms* (ed. C. B. Beck). Columbia University Press, New York.

Schwarz, J. H. (1984). What is a tarsier? In *Living fossils* (ed. N. Eldridge and S. M. Stanley), pp. 38–49. Springer Verlag, New York.

Scott, R. A., Bazghoorn, E. S., and Leopold, E. B. (1960). How old are the angiosperms? *Amer. J. Sci.* **258**-A (Bradley vol.), 284–99.

Seddon, G. (1981). Eurocentrism and Australian science: some examples. *Search* **12**, 446–50.

Shackleton, N. J. (1967). Oxygen isotope analyses and Pleistocene temperatures reassessed. *Nature, Lond.* **215**, 15–17.

Shackleton, N. J. and Opdyke, N. D. (1973). Oxygen isotopes and palaeomagnetic stratigraphy of equatorial Pacific core V28–238: Oxygen isotope temperatures and ice volumes on a 10^5 and 10^6 year scale. *Quat. Res.* **3**, 39–55.

Shackleton, N. J. and Opdyke, N. D. (1976). Oxygen-isotope and paleomagnetic stratigraphy of Pacific core V28–239, Late Pliocene to latest Pleistocene. *Mem. geol. Soc. Am.* **145**, 449–64.

Shackleton, N. J., Hall, M. A., Line, J., and Cang Shuxi (1983). Carbon isotope data in core V19–30 confirm reduced carbon dioxide concentration in the ice age atmosphere. *Nature, Lond.* **306**, 319–22.

Silver, E. A. and Moore, J. Casey (1978). The Molucca Sea collision zone, Indonesia. *J. geophys. Res.* **83**, 1681–91.

Silver, E. A., Gill, J. B., Schwartz, D., and Prasetyo, H. (1985). Evidence for a submerged and displaced continental borderland, North Banda Sea, Indonesia. *Geology* **13**, 687–91.

Simons, E. L. and Bown, T. M. (1985). *Afrotarsius chatrathi*, first tarsiiform primate (?Tarsiidae) from Africa. *Nature, Lond.* **313**, 475–7.

Simpson, G. G. (1965). *The geography of evolution*. Capricorn Books, New York.

Simpson, G. G. (1977). Too many lines; the limits of the Oriental and Australian zoogeographic regions. *Proceedings of the American Philosophical Society* **121**, 107–20.

Sims, R. W. (1978). Megadrilacea (Ologochaeta). In *Biogeography and ecology of southern Africa* vol. 2 (ed. M. J. A. Werger), pp. 661–76. Junk, The Hague.

Situmorang, B. (1982). *The formation and evolution of the Makassar Basin*. Ph.D. thesis, University of London.

Sluiter, I. R. (1984). *Palynology of Oligo-Miocene*

brown coal seams, Latrobe Valley, Victoria. Ph.D. thesis, Monash University, Clayton, Victoria.

Smith, A. C. (1963). Summary discussion on plant distribution pattern in the tropical Pacific. In *Pacific Basin biogeography* (ed. J. L. Gressit), pp. 247–9. Bishop Museum Press, Honolulu.

Smith, A. C. (1967). The presence of primitive angiosperms in the Amazon Basin and its significance in the indication of migrational routes. *Atas Simpos. Biota Amazonica* **4**, 37–59.

Smith, A. C. (1970). The Pacific as a key to flowering plant history. *Univ. Hawaii, Harold L. Lyon Arboretum Lecture* 1.

Smith, A. G., Hurley, A. M., and Briden, J.C. (1981). *Phanerozoic paleocontinental world maps*. Cambridge University Press.

Smith, J. M. B., 1977. Origins and ecology of the tropicalpine flora of Mt. Wilhelmina, New Guinea. *Biol. J. Linn. Soc.* **9**, 87–131.

Smith-White, S. (1982). Summary and re-integration. In *Evolution of the flora and fauna of arid Australia* (ed. W. R. Barker and P. J. M. Greenslade), pp. 371–9. Peacock Publications, Adelaide.

Sneath, P. H. A. and Sokal, R. R. (1973). *Principles and practice of numerical taxonomy*. Freeman, San Francisco.

Solem, A. (1979). A theory of land snail biogeographic patterns through time. In *Pathways in malacology* (ed. S. van der Spoel, A. O. van Bruggen and J. Lever), pp. 225–49. Junk, The Hague.

Solem, A. (1981). Land-snail biogeography: a true snail's pace of change. In *Vicariance biogeography* (ed. G. Nelson and D. E. Rosen), pp. 197–221. Columbia University Press, New York.

Sondaar, P. Y. (1981). The *Geochelone* faunas of the Indonesian Archipelago and their paleogeographical and biostratigraphical significance. *Mod. Quat. Res. in Southeast Asia* **6**, 111–20.

Sondaar, P. Y. (1984). Fauna evolution and mammalian biostratigraphy of Java. *Cour. Forsch. Inst. Senckenberg* **69**, 219–35.

Stamp, L. D. and Lord, L. (1923). The ecology of part of the riverine tract of Burma. *J.* Ecol.**11**. 129–59.

Stauffer, P. H. (1974). Malaya and southeast Asia in the pattern of continental drift. *Geol. Soc. Malaysia Bull.* **7**, 89–138.

Stauffer, P. H. and Gobbett, D. J. (1972). Southeast Asia as a part of Gondwanaland? *Nature Phys. Sci.* **240**, 139–40.

Stebbins, G. L. (1974). *Flowering plants: evolution above the species level*. Harvard Univ. Press, Cambridge, Mass.

Steenis, C. G. G. J. van (1934–36). On the origin of the Malaysian mountain flora. *Bull. Jard. bot. Buitenz.* Series III, **13**, 135–262; **13**, 289–417; **14**, 56–72.

Steenis, C. G. G. J. van (1949). Plumbaginaceae. *Flora Malesiana* 1, **4**, 107–12.

Steenis, C. G. G. J. van (1950). The delimitation of Malaysia and its main plant geographical divisions, *Flora Malesiana* I, **1**, lxx–lxxv.

Steenis, C. G. G. J. van (1962). The land-bridge theory in botany. *Blumea* **11**, 235–372.

Steenis, C. G. G. J. van (1965). Plant geography of the mountain flora of Mt. Kinabalu. *Proc. Roy. Soc.* B **161**, 7–38.

Steenis, C. G. G. J. van (1969). Plant speciation in Malesia with special reference to the theory of non-adaptive, saltatory evolution. *Biol. J. Linn. Soc.* **1**, 97–133.

Steenis, C. G. G. J. van (1971). *Nothofagus*, key genus to plant geography in time and space, living and fossil, ecology and phylogeny. *Blumea* **19**, 65–98.

Steenis, C. G. G. J. van (1972). *The mountain flora of Java*. E. J. Brill, Leiden.

Steenis, C. G. G. J. van (1977). Autonomous evolution in plants. Differences in plant and animal evolution. *Gard. Bull. Singapore* **29**, 103–26.

Steenis, C. G. G. J. van (1979). Plant geography of east Malesia. *Bot. J. Linn. Soc.* **79**, 97–178.

Stopes, M. C. (1912). Petrifications of the earliest European angiosperms. *Phil. Trans. R. Soc.*, Ser. B., **203**, 75–100.

Stopes, M. (1915). The Cretaceous flora. Part II. Lower Greensand (Aptian) plants of Britain. *Catalogue Mes. Plants Brit. Mus.*

Stover, L. E. and Partridge, A.D. (1973). Tertiary and Late Cretaceous spores and pollen from the Gippsland Basin, southeast Australia. *Proc. Roy. Soc. Victoria* **85**, 237-86.

Stresemann, E. (1939). Die Vögel von Celebes. *Journal für Ornithologie* **87**, 312–425.

Stuijts, I. (1984). Palynological study of Situ Bayongbong, West Java. *Mod. Quat. Res. in Southeast Asia* **8**, 17–27.

Sukamto, R. and Simandjuntak, T. O. (1983). Tectonic relationship between geologic provinces of western Sulawesi, eastern Sulawesi and Banggai-Sulu in the light of sedimentological aspects. *GRDC Bull.* **7**, 1–12.

Sun Xiang Jun, Li Ming Xing, Zhang Yi Yong, Lei Zuo Qi, Kong Zhao Chen, Li Peng, Ou Qi, and Lui Qi Na (1981). Palynology section. In *Tertiary palaeontology of north continental shelf of South China Sea*. Guandong Technology Press, Guangzhou, China.

Supriatna, S. (1980). Geologic map of the Morotai quadrangle, North Maluku, 1:250,000. Geological Research and Development Centre of Indonesia.

Takahashi, K.(1974). Palynology of the upper Aptian Tanohata Formation of the Miyako Group, northeast Japan. *Pollen et Spores* **16**, 535–64.

Takhtajan, A. (1957). On the origin of temperate flora of Eurasia. *Bot. Zh. (Leningrad)* **42**, 1635–53. (In Russian with English summary.)

Takhtajan, A. (1961). *The origin of angiospermous plants*, 2nd edn. Vysshaya Shkola, Moscow. (In Russian.)

Takhtajan, A. (1969). *Flowering plants, origin and dispersal* (trans. C. Jeffrey). Oliver and Boyd, Edinburgh; Smithsonian Institution, Washington D.C.

Takhtajan, A. (1976). Neoteny and the origin of flowering plants. In *Origin and early evolution of angiosperms* (ed. C. B. Beck), pp. 207–19. Columbia University Press, New York.

Takhtajan, A. (1983). Macroevolutionary processes in the history of plant world. *Bot. Zh. (Leningrad)* **68** (12), 1593–603. (In Russian with English summary.)

Tarling, D. H. (1972). Another Gondwanaland. *Nature, Lond.* **238**, 92–3.

Teixeira, C. (1948). *Flora mesozóica portuguesa. Part 1.* Serv. Geol. Portugal, Lisboa.

Thanikaimoni, G., Caratini, C. Venkatachala, B. S., Ramanujam, C. G. K., and Kar, R. K. (1984). Pollens d'angiospermes du Tertiare de l'Inde et leurs relations avec les pollens dur Tertiare d'Afrique. *Institut Français de Pondichery, Travaux de la Section Scientifique et Technique* **19**, 1–92.

Thenius, E. (1970). Zur Evolution und Verbreitungsgeschichte der Suidae (Artiodactyla, Mammalia). *Z. Saugetierk.* **35**, 321–42.

Thorne, R. F. (1963). Biotic distribution patterns in the tropical Pacific. In *Pacific Basin biogeography* (ed. J. L. Gressitt), pp. 311–50. Bishop Museum Press, Honolulu.

Thorne, R. F. (1976). A phylogenetic classification of angiospermae. *Evolutionary Biology*. **9**, 35–106.

Tjia, H. D., Susitno, S., Suklija, Y., Harscino, R. A. F., Rachmat, A., Hainim, J., and Djundaedi, 1984. Holocene shorelines in the Indonesian tin islands. *Mod. Quat. Res. in South East Asia* **8**, 103–17.

Torgersen, T., Jones, M. R., Stephens, A. W., Searle, D. E., and Ullman, W. J. (1985). Late Quaternary hydrological changes in the Gulf of Carpentaria. *Nature, Lond.* **313**, 785–7.

Truswell, E. M. (1983). Recycled Cretaceous and Tertiary pollen and spores in Antarctic marine sediments: a catalogue. *Palaeontographica* Abt. B. **186**, 121–74.

Truswell, E. M. (in press). The initial radiation and rise to dominance of the angiosperms. In *Rates of evolution* (ed. K. S. W. Campbell). Australian Academy of Science, Canberra.

Truswell, E. M. and Harris, W. K. (1982). The Cainozoic palaeobotanical record in arid Australia: fossil evidence for the origins of an arid-adapted flora. In *Evolution of the flora and fauna of arid Australia* (ed. W. R. Barker and P. J. M. Greenslade), pp. 67–76. Peacock Publications, Adelaide.

Uhl, N. W. (1972). Inflorescence and flower structure in *Nypa fruticans* (Palmae). *American J. Bot.* **59**, 729–43.

Uhl, N. W. and Dransfield, J. (in press). *Genera Palmarum, a classification of the palms based on the work of Harold E. Moore Jr.* The International Palm Society, Kansas.

Vail, P. R., Mitchum, R. M., and Thomson, S. (1977). Global cycles or relative changes of sea level. *Seismic stratigraphy, application to hydrocarbon exploration. Am. Assoc. Petrol. Geol. Bull. Mem.* **26**, 33–93.

Van Gorsel, H. and Troelstra, S. R. (1981). Late Neogene planktonic foraminiferal biostratigraphy and climatostratigraphy of the Solo River section (Java, Indonesia). *Marine Micropaleo.* **6**, 183–209.

Vakhrameev, V. A. (1973). Angiosperms and the boundary of the Lower and Upper Cretaceous. In *Palynology of the Mesophytic* (ed. A. F. Khlonova), pp. 131–5. Nauka, Moscow. (In Russian.)

Vakhrameev, V. A. (1981). Development of flora in the middle part of the Cretaceous Period and the ancient angiosperms. *Palaeont. J. (Moscow)* **2**, 3–14. (In Russian.)

Veevers, J. J. (1984). Australia's phanerozoic history. In *Phanerozoic earth history of Australia* (ed. J. J. Veevers). Clarendon Press, Oxford.

Verstappen, H. Th. (1975). On paleo-climates and land-form development in Malesia. *Mod. Quat. Res. in Southeast Asia* **1**, 3–55.

Walker, D. and Flenley, J. R. (1979). Late Quaternary vegetational history of the Enga District of upland Papua New Guinea. *Phil. Trans. R. Soc.* B, **286**, 265–344.

Walker, D. and Singh, G. (1981). Vegetation history. In *Australian vegetation* (ed. R. H. Groves), pp. 26–43. Cambridge University Press.

Walker, J. W. (1976*a*). Comparative pollen morphology and phylogeny of the ranalean complex. In *Origin and early evolution of angiosperms* (ed. C. B. Beck), pp. 241–99. Columbia University Press, New York.

Walker, J. W. (1976*b*). Evolutionary significance of the exine in the pollen of primitive angiosperms. In *The evolutionary significance of the exine* (ed. I. K. Ferguson and J. Muller), pp. 251–308. Academic Press, London.

Walker, J. W. and Skvarla, J. J. (1975). Primitive columellarless pollen: a new concept in evolutionary morphology of angiosperms. *Science* **187**, 445–7.

Walker, J. W. and Walker, A. G. (1984). Ultrastructure of Lower Cretaceous angiosperm pollen and the origin and early evolution of flowering plants. *Ann. Miss. Bot. Gard.* **71**, 464–521.

Walker, J. W., Brenner, G. J., and Walker, A. G. (1983). Winteraceous pollen in the Lower Cretaceous of Israel: early evidence of a magnolialean angiosperm family. *Science* **220**, 1273–5.

Wallace, A. R. (1876). *The geographical distribution of animals.* Macmillan, London.

Wallace, A. R. (1880). *The Malay Archipelago: the land of the Orang-Utan and the Bird of Paradise.* Macmillan, London.

Webb, L. J., Tracey, J. G., and Williams, W. T. (1984). A floristic framework of Australian rainforests. *Australian Journal of Ecology* **9**, 169–98.

Webb, L. J., Tracey, J. G., and Jessup, L. W. (in press). Recent evidence for autochthony of Australian tropical and subtropical rainforest floristic elements. *Telopea.*

Werren, G. L. and Sluiter, I. R. (1984). Australian rainforests—a time for reappraisal. In *Australian National Rainforests Study Report to World Wildlife Fund (Aust.), vol. 1: Proceedings of a Workshop on the Past, Present and Future of Australian Rainforests*, Griffith University, December 1983 (ed. G. L. Werren and A. P. Kershaw), pp. 488–500. Geography Department, Monash University, Melbourne.

Whitmore, T. C. (1975). *Tropical rain forests of the Far East.* Clarendon Press, Oxford.

Whitmore, T. C. (1981). Wallace's Line and some other plants. In *Wallace's Line and plate tectonics* (ed. T. C. Whitmore), Clarendon Press, Oxford.

Whitmore, T. C. and Prance, G. T. (ed.) (1987). *Biogeography and Quaternary history in tropical America.* Clarendon Press, Oxford.

Whitten, A. J., Damanik, S. J., Anwar, J. and Hisyan, N. (1984). *The ecology of Sumatra.* Gadjah Mada University Press, Yogyakarta.

Wilford, G. E., 1983. *BMR Earth Science Atlas* (ed. G. W. D'Addano and R. A. Chan). Bureau of Mineral Mineral Resources, Canberra.

Willis, J. C. (1922). *Age and area.* Cambridge University Press.

Wolfart, R., Win, U. M., Boiteau, S., Wai, U. M., Cung, U. P. U., Lwin, U. T. (1984). Stratigraphy of the Western Shan Massif, Burma. *Geologisches Jahrbuch* Reihe B, **57**, 1–95.

Wolfe, J. A., Doyle, J. A., and Page, V. M. (1975). The bases of angiosperm phylogeny: paleobotany. *Annals Missouri Bot. Gard.* **62**, 801–24.

INDEX

RANDALL LIBRARY-UNCW
3 0490 0353154 .